Anatomy &
Physiology

Therapy Basics

SECOND EDITION

Includes new material on
CELLS AND TISSUES AND REPRODUCTION

Hodder & Stoughton

A MEMBER OF THE HODDER HEADLINE GROUP

HELEN MCGUINNESS

Acknowledgements:

I would like to extend my thanks to the following people who have helped me to complete the considerable task of revising this text.
To my husband Mark for his love, support and understanding, and especially for encouraging and inspiring my work.
To my parents Roy and Val for their constant love and support, and especially to my father Roy for his skill and patience in helping to design the original illustrations.
I will always be greatly indebted to Deirdre Moynihan for her professional help and contributions throughout the preparation of the original text.
To my dear friend Dee Chase for all her support and encouragement.
To all the students and staff at the Holistic Training Centre who have been very encouraging and supportive of my creative work as an author.

This book is devoted to my precious and beautiful daughter Grace.

Cover photograph appears courtesy of Telegraph/Getty images.

A catalogue record for this title is available from the British Library

ISBN 0340802081

First published 2002

Copyright © 2002 Helen McGuinness

Typeset by Fakenham Photosetting
Printed in Great Britain for Hodder & Stoughton Educational, a division of Hodder Headline Plc, 338 Euston Road, London NW1 3BH by J.W. Arrowsmiths Ltd, Bristol.

Contents

Preface

This workbook is designed for students undertaking a course of training where knowledge of *essential* anatomy and physiology is required for safe and effective practice. The first edition was written in 1995 to assist Beauty Therapy students in learning the essential knowledge required to support their practical skills. It has now been revised and extended to cover the knowledge requirements of the growing field of Holistic Therapies.

The book in its revised format therefore forms a comprehensive framework in anatomy and physiology for those studying **Beauty Therapy at NVQ Levels II and III, Holistic / Complementary Therapies or Sports Therapy**, and for those undertaking a nursing or paramedical qualification.

The contents of the book have been written in such a way that they are not only easy to understand, but are also relevant to the therapeutic treatment application. The material in this workbook is designed to be interactive with the student, with tasks and questions for review throughout the book at intervals. Completion of the tasks and questions will help to assess overall understanding of the individual subject areas. The aim of this workbook is not only to assist candidates in learning the principles of anatomy and physiology, but also to help generate portfolio evidence of the knowledge that underpins subjects such as Beauty Therapy, Complementary Therapies and Sports Therapy. At the back of the workbook is a candidate competency record, which will enable Assessors to authenticate the evidence produced by candidates while working through the book, and may be used towards verification.

Notice to Beauty Therapy students

To assist you with learning and applying the essential knowledge required for NVQ Level 2 or 3 qualification, please refer to the study grids on pages ix/x, which indicate the knowledge relevant to the unit or units to be studied.

It was my intention in writing this workbook to provide the reader with a clear and user-friendly guide to anatomy and physiology. I hope that you will find it a valuable resource in your choice of studies.

Helen McGuinness

Study Grid – NVQ Level 3 in Beauty Therapy

ESSENTIAL KNOWLEDGE & UNDERSTANDING	Unit 11	Unit 12	Unit 13	Unit 14	Unit 15	Unit 16a	Unit 16b	Unit 17	Unit 18	Unit 19	Unit 20	Unit 21 (Add't)	Unit 22 (Add't)
Chapter 2 – The Skin				●	●	●	●	●			●		
Chapter 3 – The Hair						●	●						
Chapter 4 – The Nail										●			
Chapter 5 – The Skeletal System				●	●			●			●		
Chapter 6 – Joints													
Chapter 7 – The Muscular System				●Body	●			●			●Face		
Chapter 8 – The Circulatory System				●	●	●	●	●			●		
Chapter 9 – The Lymphatic System				●Body	●	●	●	●			●Face		
Chapter 11 – The Respiratory System													
Chapter 12 – The Olfactory System								●					
Chapter 13 – The Nervous System				●	●			●			●		
Chapter 14 – The Endocrine System						●	●						
Chapter 16 – The Female Breast													
Chapter 17 – The Digestive System													
Chapter 18 – The Urinary System													

Study Grid – NVQ Level 2 in Beauty Therapy

	Unit 1	Unit 2	Unit 3	Unit 4	Unit 5	Unit 6	Unit 7	Unit 8	Unit 9	Unit 10
Essential Knowledge & Understanding										
Chapter 2 – The Skin					●	●			●	
Chapter 3 – The Hair							●			
Chapter 4 – The Nail									●	
Chapter 5 – The Skeletal System (head, face, neck and shoulder girdle bones)						●				
Chapter 7 – The Muscular System (facial, neck and shoulder muscles)						●				
Chapter 8 – The Circulatory System (head and neck)						●				
Chapter 9 – The Lymphatic System (head and neck)						●				

CHAPTER 1

Cells and Tissues

In any form of therapeutic treatment application, the prime concern must be for the health and well being of the client. Therapists must therefore have a sufficient degree of knowledge of anatomy and physiology, in order to be able to ensure safety and efficacy in therapeutic treatments.

The most effective therapeutic application depends on the therapist's ability to locate regions of the body and an understanding of the structures being treated. Whilst this may involve essentially practical skills and techniques, knowledge of anatomy and physiology is a necessary prerequisite to be able to apply the skills effectively.

Therapists are not required to have as much in-depth knowledge as a medical professional, because they are not involved in diagnosis. However, they should have a sound knowledge of anatomy and physiology to understand how the body works, and how different parts of the body integrate with one another.

With this foundation, a therapist will then have an understanding of how common diseases affect specific functions, and should be able to recognise conditions where treatment may be beneficial or potentially detrimental to the client's health and well being. This chapter introduces some basic terms that underpin anatomy and physiology, and gives an overview of the organisation of the body from both a cellular and a structural level.

By the end of the chapter you will be able to recall and relate the following knowledge to your practical work:

▶ definitions of the terms anatomy, physiology, homeostasis and metabolism
▶ the parts of a cell's structure and their functional significance
▶ definitions of cellular processes such as diffusion, osmosis and active transport
▶ the four main tissue types and their individual classifications.

Definition Of terms

Anatomy and Physiology

The terms anatomy and physiology are two very distinct but interrelated terms, that combine to form the basis of the body's operation as a whole. **Anatomy** is the scientific study of the structures of the body and the relationship of its parts, whilst **physiology** is the scientific study of the way in which the body works.

Anatomy and physiology are inseparable in that structure is determined by function, and function is carried out through structure. For example, anatomy would describe the position of a muscle, while physiology would describe how the muscle contracts.

KEY NOTE

The best way to view anatomy and physiology is from a holistic viewpoint, and while it may be easier to learn the subject matter by dividing the body into individual systems and organs, it has to be appreciated that the workings of the body as a whole are highly complex and integrated.

Homeostasis

Traditionally, the body is divided into different systems according to their specific functions. However, the ultimate purpose of each system is to maintain a constant internal environment for each cell to enable it to survive.

The human body is exposed to a constantly changing external environment. These changes are neutralised by the internal environment of blood, lymph and tissue fluids that bathe and protect the cells. Body parts function efficiently *only* when the concentrations of water, food substances, oxygen and the conditions of heat and pressure remain within certain narrow limits. The process by which the body maintains a stable internal environment for its cells and tissues is called **homeostasis**.

The body and its systems are constructed in such a way that all systems synergistically with one another with one overall aim – to maintain hom Examples of homeostatic mechanisms in the body include the regulation of body temperature, blood pressure and blood sugar levels.

Metabolism

Homeostasis is maintained by adjusting the **metabolism** of the body. This is the term used to describe the physiological processes that take place in our bodies, to convert the food we eat and the air we breathe, into the energy we need to function. Metabolism is essentially the basic working of the body cells.

The rate at which a person consumes energy in activities and body processes is known as the **metabolic rate.** The minimum energy required to keep the body alive is known as the **basal metabolic rate.**

Levels of Organisation in the Body

The structure of the human body involves five principal levels of organisation:

- **Atoms and molecules** – these represent the lowest level of organisational complexity in the body. At the chemical level, the smallest unit of matter is the atom.
- **Cells** – are the fundamental structural and functional units of life, and are the smallest units that show characteristics of life.
- **Tissue** – this is a group of similar cells that perform a certain function (for example nervous and muscular tissues).
- **Organs** – tissues are grouped into structurally and functionally integrated units called organs (for example, the lungs or the heart).
- **Systems** – a system is a group of organs that work together to perform specific functions. The systems of the body include the circulatory, the skeletal, the integumentary (the skin), the respiratory, the reproductive, the muscular, the endocrine, the nervous, the urinary and the digestive systems.

KEY NOTE

Although a therapist's work is primarily at the systemic level, the benefits of therapeutic treatments are due to chemical changes produced at the cellular level. It is therefore important for a therapist to study the cells and tissues of the body in order to understand the effects of therapeutic treatment application.

Cells

The cell is the fundamental unit of all living organisms and is the simplest form of life that can exist as a self-sustaining unit. Cells are therefore the building blocks of the human body.

Cells in the body take many forms, the size and shape being largely dependent on their specialised function. For example, different cells help fight disease, transport oxygen, produce movement, manufacture proteins and chemicals, and store nutrients.

Cells consist of four elements: carbon, oxygen, hydrogen and nitrogen, plus trace elements such as iron, sodium and potassium. Trace elements are of significance in certain cellular functions, for example calcium is needed for blood clotting. Besides the four primary elements, water makes up 60–80 per cent of all cells.

Cell Organelles

Molecules combine in very specific ways to form **cell organelles**, the basic structures found in cells. The basic component parts of the cell are called organelles and each has a particular functional significance within the cell that allows it to live.

Typical cell organelles include the:

- cell membrane
- cytoplasm
- nucleus
- centrioles
- ribosomes
- endoplasmic reticulum
- mitochondria
- lysosome
- golgi body

Despite the great variety of cells in the body, they all have the same basic structure.

Cell Membrane

The cell membrane is a fine membrane that encloses the cell and protects its contents. It is called semi-permeable because it selectively controls the inward and outward movement of molecules into and out of the cell. Oxygen, nutrients, hormones and proteins are taken into the cell as needed, and cellular waste such as carbon dioxide passes out through the membrane.

- As well as governing the exchange of nutrients and waste materials, its function is also to maintain the shape of the cell.

Cytoplasm

The cytoplasm is the gel-like substance that is enclosed by the cell membrane. The cytoplasm contains the nucleus and the small cellular structures called organelles. Most cellular metabolism occurs within the cytoplasm of the cell.

Nucleus

The nucleus is the largest organelle in the cytoplasm. It is the control centre of the cell, regulating the cell's functions and directing nearly all metabolic activities. The nucleus governs the specialised work performed by the cell and the cell's own growth, repair and reproduction.

- All cells have at least one nucleus at some time in their existence.
- The nucleus is significant in that it contains all the information required for the cell to function, and it also controls all cellular operations.
- The information required by the cell is stored in DNA (deoxyribonucleic acid), which carries the genetic materials for replication of identical molecules. The DNA strands are found in threadlike structures known as chromosomes. Each human cell has 23 pairs of chromosomes.
- Inside the nucleus is the **nucleolus** which contains ribonucleic acid (RNA) structures that form ribosomes.
- The nucleus is surrounded by a perforated outer membrane called the **nuclear membrane**; materials move across it to and from the cytoplasm.

Centrioles

Contained within the centrosomes (an area of clear cytoplasm found next to the nucleus) are the small spherical structures called centrioles. They are associated with cell division, or mitosis.

- During cell division, the centriole divides in two and migrates to opposite sides of the nucleus to form the **spindle poles**.

Ribosomes

Ribosomes are tiny organelles made up of ribonucleic acid (RNA) and protein. They may be fixed to the walls of the endoplasmic reticulum (known as rough ER), or may float freely in the cytoplasm.

- Their function is to manufacture proteins for use within the cell, and also to produce other proteins that are exported outside the cell.

Endoplasmic Reticulum

This is a series of membranes continuous with the cell membrane. It allows movement of materials from one part of the cell to another, and can be thought of as an intracellular transport system. It links the cell membrane with the nuclear membrane and therefore assists movement and materials out of the cell.

▶ It contains enzymes and participates in the synthesis of proteins, carbohydrates and lipids.
▶ The endoplasmic reticulum stores material, transports substances inside the cell, and detoxifies harmful agents.
▶ Some endoplasmic reticula appear smooth, while others appear rough due to the presence of ribosomes.

Mitochondria

Mitochondria are oval-shaped organelles that lie in varying numbers within the cytoplasm. The mitochondria are the site of the cell's energy production. The mitochondria provide most of a cell's ATP (adenosine triphosphate): this is a compound that stores the energy needed by the cell.

▶ The work of the mitochondria is assisted by enzymes, which are proteins that speed up chemical changes.
▶ The mitochondria are therefore responsible for providing the energy which powers the cell's activities.

Lysosome

These are round sacs present in the cytoplasm. They contain powerful enzymes which are capable of digesting proteins. Their function is to destroy any part of the cell that is worn out so that it can be eliminated – this is known as lysis.

Golgi body / apparatus

This is a collection of flattened sacs within the cytoplasm. The golgi apparatus is typically located near the nucleus and attached to the endoplasmic reticulum. It is the 'packaging department' of the cell, as it stores the protein manufactured in the endoplasmic reticulum and later transports it out of the cell.

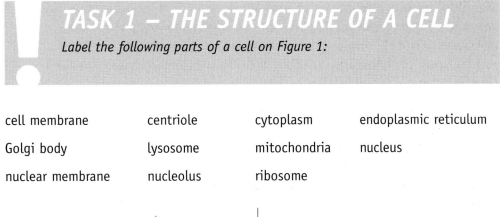

TASK 1 – THE STRUCTURE OF A CELL

Label the following parts of a cell on Figure 1:

cell membrane   centriole   cytoplasm   endoplasmic reticulum

Golgi body   lysosome   mitochondria   nucleus

nuclear membrane   nucleolus   ribosome

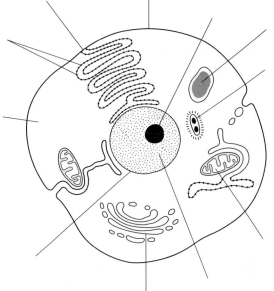

Figure 1
The structure of a cell

Functions of Cells

In order for a cell to survive, it must be able to carry out a variety of functions.

Respiration

Every cell requires oxygen for the process of metabolism. Oxygen is absorbed through the cell's semi-permeable membrane and is used to oxidise nutrient material to provide heat and energy.

The waste products produced as a result of cell respiration include carbon dioxide and water. These are passed out from the cell through its semi-permeable membrane.

Growth

Cells have the ability to grow until they are mature and ready to produce. A cell can grow and repair itself by manufacturing protein.

Excretion

During metabolism, various substances are produced which are of no further use to the cell. These waste products are removed through the cell's semi-permeable membrane.

Movement

Movement may occur in the whole or part of a cell. White blood cells, for example, are able to move freely.

Irritability

A cell has the ability to respond to a stimulus, which may be physical, chemical or thermal. For example, a muscle fibre contracts when stimulated by a nerve cell.

Reproduction

When growth is complete in a cell, reproduction takes place. The cells of the human body reproduce, or divide by the process of mitosis.

Cellular Processes

In order to function properly, a cell must maintain a stable internal environment. The transport of materials therefore has to be achieved without an excessive build-up of chemicals.

When certain molecules are needed, such as glucose, the cell will take these in and discard other materials in order to preserve equilibrium.

Diffusion

As chemicals become concentrated outside the cell, small molecules flow through the cell membrane until a balance exists. The process in which small molecules move from areas of *high* concentration to those of *lower* concentration is called **diffusion**.

▶ Diffusion is the means by which the cells lining the small intestines take in digestive products to be utilised by the body.

Osmosis

This process refers to the movement of water through the cell membrane from areas of *low* chemical concentration to areas of *high* chemical concentration.

▶ Osmosis allows for the dilution of chemicals, which are unable to cross the cell membrane by diffusion, in order to maintain equilibrium within the cell.

Active Transport

This is the process, using chemical energy, by which the cell takes in larger molecules that would be otherwise unable to enter in sufficient quantities. Carrier molecules within the cell membrane bind themselves to the incoming molecules, rotate around them and release them into the cell. This is the means by which the cell absorbs glucose.

The Cell's Life Cycle

Cells undergo many divisions from the time of fertilisation to physical maturity. When a single cell undergoes division, it forms two daughter cells that are identical to the original cell. A cell may live from a few days to many years, depending on its type.

Cells divide in two ways: mitosis and meiosis.

▶ **Mitosis** is when a single cell produces two genetically identical daughter cells. Mitosis is the way in which new body cells are produced for both growth and repair. Division of the nucleus takes place in four phases (prophase, metaphase, anaphase and telophase) and is followed by the division of the cytoplasm to form the daughter cells. See figure 2 on page 10.
▶ **Meiosis** can be seen in the testis and ovary during the formation of sperm and ova in sexual reproduction. It is a type of cell division that produces four daughter cells, each having half the number of chromosomes of the original cell. It occurs before the formation of sperm and ova, and the normal number of chromosomes is restored after fertilisation.

Tissues

Due to the complexity of the human body, it is not possible for every cell to carry out all the functions required by the body; some cells therefore become specialised to form a group of cells or tissues. The cells of a tissue are embedded in or

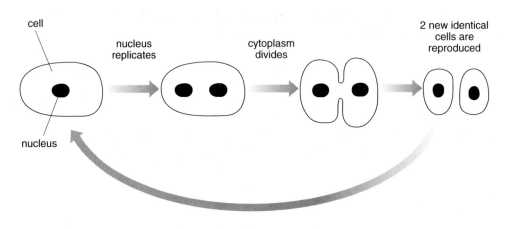

Figure 2
Mitosis

surrounded by nonliving material called the **matrix**; the amount and configuration being dependent on the type of tissue and the support it needs.

Tissues are defined as a group of cells that act together to perform a specific function. There are four major types of tissues in the human body:

▶ **epithelial**
▶ **connective**
▶ **muscle**
▶ **nervous**.

All four types of tissue have special purposes, and therefore have varying different rates of cellular regeneration:

▶ Epithelial tissue is constantly being renewed by the process of cell division or mitosis.
▶ Bone tissue and adipose connective tissue are highly vascular and therefore heal quickly.
▶ Muscle tissue takes longer to regenerate.
▶ Nervous tissue regenerates very slowly.

The less vascular forms of connective tissue such as ligaments and tendons are even slower to heal than muscle tissue, and cartilage is amongst the slowest to heal.

Tissues organise into **organs**. Organs are defined as a group of two or more tissues that act together to perform a specific function.

> ### KEY NOTE
>
> It is important for a therapist to have a basic understanding of the tissues in order to understand the structure and functions of the body's organs.

Epithelial Tissue

Epithelial tissue consists of sheets of cells which cover and protect the external and internal surfaces of the body, and line the inside of hollow structures. They specialise in moving substances in and out of the blood during secretion, absorption and excretion.

Because they are subject to a considerable amount of wear and tear, epithelial cells reproduce very actively. Usually there is little matrix present in epithelial tissues. The matrix present tends to form continuous sheets of cells, with the cells held very close together. A thin permeable basement membrane attaches epithelial tissues to the underlying connective tissue.

Epithelial tissue, which consists of cells closely packed together, comes in various shapes. There are two categories of epithelial tissue:

▶ simple (single layered), or
▶ compound (multi-layered).

Simple Epithelium

Simple epithelium have only one layer of cells over a basement membrane. Being thin, they are fragile, and are found only in areas inside the body which are relatively protected – such as the lining of the heart and blood vessels, and the lining of body cavities. They also line the digestive tract and in the exchange surfaces of the lungs, where their thinness is an advantage for speedy absorption across them.

There are four different types of simple epithelium, named according to their shape and functions they perform:

1   squamous

2 cuboidal
3 columnar
4 ciliated.

Squamous Epithelium

These cells are flat, scale-like cells with a central nucleus. The cells fit closely together, rather like a pavement, producing a very smooth surface. As it is so flat, this type of tissue is particularly well suited to cellular processes such as diffusion.

▶ Squamous epithelium lines the alveoli of the lungs. It also lines blood vessels and the heart, where it is known as endothelium.

Cuboidal Epithelium

The cube-like shape of these cells is only visible when the tissue is sectioned through at right angles.

▶ Cuboidal epithelium is found in areas where absorption or secretion takes place, such as the ovaries, kidney tubules, the thyroid gland, the pancreas and the salivary glands.

Columnar Epithelium

These cells are of much greater height than width. They consist of a single layer of cells of cylindrical shape, with the nucleus being situated towards the base of the cell.

▶ This type of tissue lines the small and large intestine, the stomach and the gall bladder, and is involved in secretion and absorption.
▶ Its thickness helps to protect underlying tissues and many of these types of cells are modified for a particular function. For example, in the small intestine the plasma membrane is folded into microvilli, whose function is the absorption of nutrients.

Ciliated Epithelium

This is a form of columnar epithelium and has hair-like projections called cilia from its surface.

▶ This type of cell lines the respiratory system, and the cilia carry unwanted particles along with mucus out of the system, maintaining its cleanliness. It also lines the uterine tubes to help propel the ova towards the uterus.

Compound Epithelium

The main function of compound epithelium is to protect underlying structures. Compound epithelium contains two or more layers of cells. There are two main types:

1 stratified
2 transitional.

Stratified Epithelium

This is composed of a number of layers of cells of different shapes. In the deeper layers the cells are mainly columnar in shape, and as they grow towards the surface they become flattened.

There are two types of stratified epithelium: keratinised, and non-keratinised epithelium.

▶ **Non-keratinised stratified epithelium** is found on wet surfaces that may be subject to wear and tear; for example, the conjunctiva of the eyes, the lining of the mouth, the pharynx and the oesophagus.
▶ **Keratinised stratified epithelium** is found on dry surfaces, such as lining the skin, hair and nails. The surface layers of keratinised cells are dead cells – they are continually being rubbed off and replaced from below. They give protection and prevent drying out of the cells, in the deeper layers from which they develop.

Transitional Epithelium

This is composed of several layers of pear-shaped cells which change shape when they are stretched. This type of tissue lines the uterus, bladder and the pelvis of the kidney.

Connective Tissue

Connective tissue is the most abundant type of tissue in the body. It connects tissues and organs by binding the various parts of the body together, and helps to give protection and support. Connective tissues are made up of cells and matrix.

Connective tissue cells are often more widely separated from each other than those forming epithelial tissue, and the space between cells is larger and is filled with a large amount of nonliving matrix. There may or may not be fibres in the matrix, which may be either a semi-solid jelly-like consistency, or dense and rigid, depending on the position and function of the tissue.

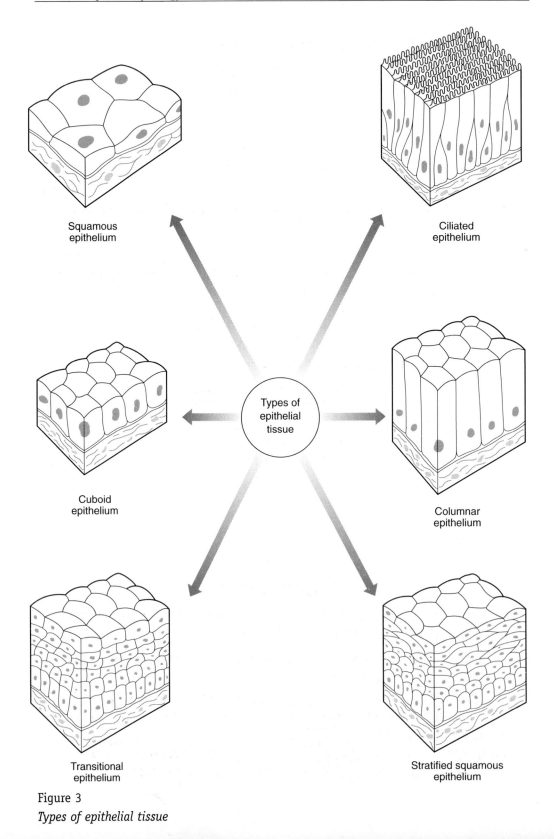

Squamous
epithelium

Ciliated
epithelium

Cuboid
epithelium

Types of
epithelial
tissue

Columnar
epithelium

Transitional
epithelium

Stratified squamous
epithelium

Figure 3
Types of epithelial tissue

There are several different types of connective tissue:

- areolar
- adipose
- white fibrous
- yellow elastic

- lymphoid
- blood
- bone
- cartilage.

Areolar Tissue

This is the most widely distributed type of connective tissue in the body. This type of tissue is composed of cells called **fibrocytes**, which are widely separated by white and reticular fibres, as well as yellow elastic fibres.

- This tissue allows for elasticity and is found in almost every part of the body connecting and supporting organs: under the skin, between muscles, supporting blood vessels and nerves, and in the alimentary canal.

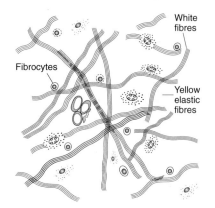

Figure 4
Areolar tissue

Adipose Tissue

This tissue is composed of specialised cells for the storage of fat – **adipocytes**. These cells are present within the matrix, but few fibres are present.

- Its function is to provide protection to the organs close to it, to reduce heat loss and to act as an emergency energy reserve.
- This type of tissue supports organs such as the kidneys and the eyes, and is found between bundles of muscle fibres, in the yellow bone marrow of long bones, and as a padding around joints.
- Along with areolar tissue, adipose tissue is found *under* the skin, in the subcutaneous layer, which gives the body a smooth, continuous line.

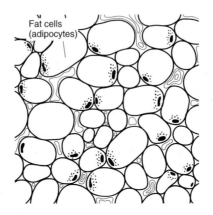

Fat cells (adipocytes)

Figure 5
Adipose tissue

White Fibrous Tissue

This is a strong, connecting tissue made up of mainly closely packed bundles of white, collagenous fibres, with very little matrix.

▶ Cells called **fibrocytes** are present between the bundles. This type of connective tissue forms tendons (which attach muscle to bone), ligaments which tie bones together, and as an outer protective covering for some organs such as the kidney and the bladder.
▶ The function of this tissue is to provide strong attachment between different structures.

Yellow Elastic Tissue

This consists of branching yellow elastic fibres with fibrocytes in the spaces between the fibres.

▶ It is found in areas where alteration of shape is needed, such as the arteries, trachea and bronchi and the lungs.
▶ Its function is to allow stretching of various organs, followed by a return to their original shape and size.

Lymphoid Tissue

This tissue has a semi-solid matrix with fine branching fibres. The cells contained within this tissue are specialised and are called **lymphocytes**.

▶ It is found in the lymph nodes, the spleen, the tonsils, the adenoids, walls of the large intestine, and glands in the small intestine.

▶ This type of tissue forms part of the lymphatic system whose function is to protect the body from infection.

Blood (fluid or liquid tissue)

Blood is also known as liquid connective tissue. It contains the blood cells **erythrocytes**, **leucocytes** and **thrombocytes**, which float within a fluid called **plasma**. Blood helps maintain homeostasis of the body by transporting substances throughout the body, by resisting infection and by maintaining heat.

Bone Tissue

Bone is the hardest and most solid of all connective tissues. It consists of:

▶ compact tissue – a dense type designed for strength
▶ spongy cancellous tissue for structure bearing
▶ collagenous fibres for strength and mineral salts for hardness.

Bone is infused with many blood vessels and nerves through a membranous sheath surrounding the bone called the **periosteum**.

For additional information on bone tissue and its cellular development see Chapter 5.

Cartilage

This is a much firmer tissue than any of the other connective tissues and the matrix is quite solid. For descriptive purposes, cartilage is divided into three types:

▶ Hyaline cartilage
▶ White fibrous cartilage
▶ Yellow elastic fibrocartilage.

Hyaline Cartilage

This is a smooth, bluish-white, glossy tissue. It contains numerous cells called **chondrocytes** from which cartilage is produced.

▶ Hyaline cartilage is the most abundant type. It is found on the surfaces of the parts of bones which form joints, and forms the costal cartilage which attaches the ribs to the sternum, part of the larynx, trachea and bronchi.
▶ Hyaline cartilage provides a hard-wearing, low friction surface within joints and is flexible to provide support in, for example, the nose and the trachea.

White Fibrous Cartilage

This is composed of bundles of collagenous white fibres, with chondrocytes scattered amongst them. It has a solid matrix, and is tough, but slightly flexible.

▶ It is found as pads between the bodies of the vertebrae called the intervertebral discs, and in the symphysis pubis which joins the pubis bones together.
▶ Its function is one of support and to join together or fuse certain bones.

Yellow Elastic Fibrocartilage

This cartilage consists of yellow elastic fibres running through a solid matrix, between which chondrocytes are situated.

▶ It forms the pinna (lobe of the ear) and the epiglottis.
▶ The function of this type of cartilage is to provide support and to maintain shape.

Muscle Tissue

Muscle tissue is very elastic, and therefore has the unique ability to provide movement by shortening as a result of contraction. Muscle tissue is made up of contractile fibres, usually arranged in bundles and surrounded by connective tissue.

There are three types of muscle tissue: voluntary (skeletal), involuntary or smooth, and cardiac muscle. The types of muscle tissue are discussed in more detail in Chapter 7 – The Muscular System.

Nervous Tissue

Nervous tissue consists of cells called neurones which can pick up and transmit electrical signals by converting stimuli into nerve impulses. Nervous tissue has the characteristics of excitability and conductivity, and its functions are to coordinate and regulate body activity. Nervous tissue and neurones are discussed in more detail in Chapter 13 – The Nervous System.

Membranes

Membranes are thin, soft, sheet-like layers of tissue that cover a cell, an organ or a structure. They line tubes or cavities, and divide and separate one part of a cavity from another. There are three basic types of membranes in the body:

1 Mucous membrane
2 Serous membrane
3 Synovial membrane

Mucous and serous membrane are composed of epithelial tissue, while synovial membrane is composed of various types of connective tissue.

Mucous Membrane

Mucous membrane lines openings to the external environment, such as the respiratory, digestive, urinary and reproductive tracts. Mucous membrane secretes a viscous slippery fluid called mucus that coats and protects the underlying cells.

Serous Membrane

Serous membrane lines body cavities that are not open to the external environment, and cover many of the organs. These membranes consist of two layers: a parietal layer, which lines the wall of body cavities, and a visceral layer, which provides an external covering to organs in body cavities.

▶ They secrete a thin, watery fluid that lubricates organs to reduce friction as they rub againse one another and against the wall of the cavities.

▶ Examples of serous membranes include the pericardium of the heart, the pleural membranes in the lungs and the peritoneum lining the abdominal organs.

Synovial Membrane

Synovial membrane lines the joint cavities of freely movable joints such as the shoulder, hip and knee. These membranes secrete synovial fluid that provides nutrition and lubrication to the joint so that it can move freely without undue friction.

▶ This type of membrane is also found in **bursae**, which are protective sacs located around a joint cavity; between layers of muscle and connective tissue; and wherever the body needs extra protection

TASK 2 – TYPES OF TISSUE

Identify the types of tissue from the descriptions given.

Type of Tissue	Description
	This type of tissue is made up of white collagenous fibres, and forms tendons and ligaments.
	This type of tissue acts as an emergency energy reserve, and is made up of fat cells.
	This type of tissue is the most widely distributed type of tissue in the body, and is composed of cells called fibrocytes, as well as white, reticular and yellow elastic fibres. It is found in almost every part of the body connecting and supporting organs.
	This tissue is found where an alteration of shape is needed, and allows organs such as the lungs to stretch and then return to their original size.

SELF ASSESSMENT QUESTIONS – CELLS AND TISSUES

1. State the difference between the following terms:

a) anatomy
b) physiology

2. What is a cell?

3. State the significance of the following parts of a cell's structure:

a) cell membrane

....

....

b) nucleus

....

....

c) ribosome

....

....

d) golgi body

....

....

4. Identify the following:

a) the organelle that is able to digest any worn out part of the cell

....

....

b) the organelles that are associated with cell division

....

....

c) the organelle that powers the cell activities

....

....

5. Briefly define the following terms:

a) homeostasis

b) cell metabolism

6. State the difference between the following two cellular processes:

a) diffusion
b) osmosis

7. Briefly describe the process of cell division or mitosis.

8. State three functions of cells.

9. State two examples of where the following types of tissues may be found in the body:

a) squamous epithelium

b) ciliated epithelium

c) keratinised stratified epithelium

10. Briefly state where the following may be found, stating their function/s in each case:

a) hyaline cartilage

....

....

b) mucous membrane

....

....

c) synovial membrane

....

....

CHAPTER 2

The Skin

The skin is a very important organ to a therapist as it represents the common foundation for all practical treatments carried out in the salon; therefore an understanding of its structure and functions is essential for carrying out treatments safely and effectively.

A competent therapist needs to be able to:

▶ relate knowledge of the skin to the physical effects of treatments.

By the end of this chapter you will be able to relate the following knowledge to your practical work carried out in the salon:

▶ the structure and function of the epidermis
▶ the structure and function of the dermis
▶ the structure and function of the subcutaneous layer
▶ the appendages of the skin and their functional significance
▶ the blood, lymphatic and nerve supply to the skin
▶ the functions of the skin
▶ treatable and non-treatable skin diseases and disorders.

No other body system is more easily exposed to infections, disease, pollution or injury than the skin; yet no other body system is as strong and resilient.

Before looking at the structure of the skin, let's consider a few facts . . .

▶ The skin is a very large organ covering the whole body.

▶ It varies in thickness on different parts of the body. It is thinnest on the lips and eyelids, which must be light and flexible, and thickest on the soles of the feet and palms of the hands where friction is needed for gripping.

▶ As the skin is the external covering of the body, it can be easily irritated and damaged and certain symptoms of disease and disorders may occur.

▶ Each client's skin varies in colour, texture, and sensitivity and it is these individual characteristics that make each client unique.

The appearance of the skin reflects a client's physiology. Observation of a client's skin will indicate their nutrition, circulation, age, immunity, genetics, as well as environmental factors, which all play a significant role in the skin's colour, condition and tone.

Let's take a closer look at the structure of the skin. There are two main layers of the skin:

▶ The **epidermis** which is the outer thinner layer
▶ The **dermis** which is the inner thicker layer

Below the dermis is the **subcutaneous** layer which attaches to underlying organs and tissues.

The Epidermis

The epidermis is the most superficial layer of the skin and consists of five layers of cells. The three outermost layers consist of dead cells as a result of the process of keratinisation; the cells in the very outermost layer are dead and scaly and are constantly being rubbed away by friction.

KEY NOTE

Keratin is the tough fibrous protein found in the epidermis, the hair and the nails. The keratin found in the skin is constantly being shed. Keratinisation refers to the process that cells undergo when they change from living cells with a nucleus (which is essential for growth and reproduction) to dead, horny cells without a nucleus. Cells which have undergone keratinisation are therefore dead. Keratin has a protective function in the skin as the keratinised cells form a waterproof covering, helping to stop the penetration of bacteria and protect the body from minor injury.

The inner two layers are composed of living cells. The epidermis does not have a system of blood vessels, only a few nerve endings which are present in the lower epidermis. Therefore, all nutrients pass to the cells in the epidermis from blood vessels in the deeper dermis. The five layers of cells of the epidermis are as follows:

- the basal cell layer
- the prickle cell layer
- the granular layer
- the clear layer
- the horny layer

Basal Cell Layer

This is the deepest of the five layers. It consists of a single layer of column cells on a basement membrane which separates the epidermis from the dermis. In this layer, the new epidermal cells are constantly being reproduced. These cells last about six weeks from reproduction or **mitosis** before being discarded into the horny layer. New cells are therefore formed by division, pushing adjacent cells towards the skin's surface. At intervals between the column cells, which divide to reproduce, are the large star-shaped cells called melanocytes, which form the pigment melanin, the skin's main colouring agent.

KEY NOTE

Melanin is produced by special cells called **melanocytes**, which are found in the basal cell layer of the epidermis. Melanocytes have finger-like projections which are capable of injecting melanin granules into neighbouring cells of the epidermis. This explains why melanin is found in the basal layer, prickle cell layer and granular layer of the epidermis. Melanin is responsible for the colour of the skin and hair and helps protect the deeper layers of the skin from the damaging effects of ultra violet radiation.

Prickle Cell Layer

This is known as the prickle cell layer because each of the rounded cells contained within it have short projections which make contact with the neighbouring cells and give them a prickly appearance. The living cells of this layer are capable of dividing by the process mitosis.

Granular Layer

This layer consists of distinctly shaped cells, containing a number of granules which are involved in the hardening of the cells by the process keratinisation. This layer links the living cells of the epidermis to the dead cells above.

Clear Layer

This layer consists of transparent cells which permit light to pass through. It consists of three or four rows of flat dead cells, which are completely filled with keratin; they have no nuclei as the cells have undergone mitosis. The clear layer is very shallow in facial skin but thick on the soles of the feet and palms of the hands and is generally absent in hairy skin.

Horny Layer

This is the most superficial, outer layer, consisting of dead, flattened, keratinised cells which have taken approximately a month to travel from the germinating layer. This outer layer of dead cells is continually being shed: this process is known as desquamation.

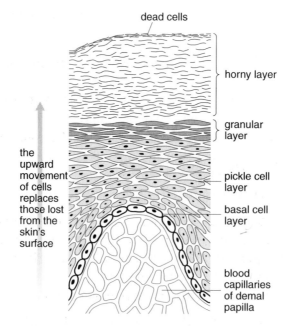

Figure 6
Cell regeneration of the epidermis

Cell regeneration occurs in the epidermis by the process mitosis. It takes approximately a month for a new cell to complete its journey from the basal cell layer where it is reproduced to the granular layer, where it becomes keratinised to the horny layer where it is desquamated.

The Dermis

The dermis is the deeper layer of the skin, and its key functions are to provide support, strength and elasticity. The dermis has a superficial **papillary** layer and a deep **reticular** layer. The superficial papillary layer is made up of adipose connective tissue and is connected to the underside of the epidermis by cone-shaped projections called dermal papillae, which contain nerve endings and a network of blood and lymphatic capillaries. The many dermal papillae of the papillary layer form indentations in the overlying epidermis, giving it an irregular appearance.

KEY NOTE

The key function of the papillary layer of the dermis is to provide vital nourishment to the living layers of the epidermis above.

The deep reticular layer is formed of tough fibrous connective tissue which contains the following:

- **collagen** fibres containing the protein collagen, which gives the skin its strength and resilience
- **elastic** fibres containing a protein called elastin, which gives the skin its elasticity
- **reticular** fibres which help to support and hold all structures in place.

These fibres all help maintain the skin's tone.

Cells present in the dermis are as follows:

- **mast** cells which secrete histamine, causing dilation of blood vessels to bring blood to the area. This occurs when the skin is damaged or during an allergic reaction
- **phagocytic** cells, which are white blood cells that are able to travel around the dermis destroying foreign matter and bacteria
- **fibroblasts**, which are cells that form new fibrous tissue.

KEY NOTE

The principal function of the dermis is to provide nourishment to the epidermis and to give a supporting framework to the tissues.

Appendages of the Skin

The appendages are accessory structures that lie in the dermis of the skin and project onto the surface through the epidermis. These include the hair, the erector pili muscle, sweat and sebaceous glands.

Hair

Hair is an appendage of the skin which grows from a sac-like depression in the epidermis called a hair follicle. It grows all over the body, with the exception of the palms of the hands and the soles of the feet. Its primary functions are to protect the skin, help keep us warm and to assist in our sense of touch.

The hair is composed mainly of the protein keratin and is therefore a dead structure.

Longitudinally the hair is divided into three parts: the hair shaft, hair root and hair bulb. Internally the hair has three layers (cuticle, cortex and medulla) which all develop from the matrix, which is the active growing part of the hair.

Hair Follicle

The hair follicle is a pore-like indentation in the epidermis that extends down to the dermis and surrounds the root of the hair.

The hair follicle is a sac or sheath of epidermal cells and connective tissue which surrounds the root of the hair.

The function of the hair follicle is to hold the hair in place and enclose the hair shaft. It also provides the hair with a vital source of nourishment, as it contains the dermal papilla which supplies blood to the hair.

Erector Pili Muscle

This is a small smooth muscle, made up of sensory fibres. It is attached at an angle to the base of a hair follicle, which serves to make the hair stand erect in response to cold, or when experiencing emotions such as fright and anxiety.

Erector pili muscles are the weak muscles associated with hairs.

Sweat Glands

There are two types of sweat glands – eccrine and apocrine. The majority are **eccrine** glands, which are simple coiled tubular glands that open directly onto the surface of the skin. There are several million of them distributed over the surface of the skin, although they are most numerous in the palms of the hands and the soles of the feet.

- Their function is to regulate body temperature and help eliminate waste products.
- Their active secretion sweat is under the control of the sympathetic nervous system.
- Heat-induced sweating tends to begin on the forehead and then spreads to the rest of the body, whereas emotionally-induced sweating, stimulated by fright, embarrassment or anxiety begins on the palms of the hands and in the axilla, and then spreads to the rest of the body.

Apocrine glands are connected with hair follicles and are only found in the genital and underarm regions. They produce a fatty secretion; breakdown of the secretion by bacteria leads to body odour.

Sebaceous Glands

These glands are small sac-like pouches found all over the body, except for the soles of the feet and the palms of the hands. They are more numerous on the scalp, face, chest and back.

Sebaceous glands commonly open into a hair follicle but some open onto the skin surface. They produce an oily substance called sebum which contains fats, cholesterol and cellular debris.

- Sebum is mildly antibacterial and antifungal.
- The secretion of sebum is stimulated by the release of hormones, primarily androgens.
- Sebum coats the surface of the skin and the hair shafts where it prevents excess water loss, lubricates and softens the horny layer of the epidermis, and softens the hair.

Blood Supply

Unlike the epidermis, the dermis has an abundant supply of blood vessels which run through the dermis and the subcutaneous layer of the skin. Arteries carry oxygenated blood to the skin via arterioles (small arteries) and these enter the dermis from below and branch into a network of capillaries around active or growing structures.

These capillary networks form in the dermal papillae to provide the basal cell layer of the epidermis with food and oxygen. The networks also surround two appendages of the skin: the sweat glands and the erector pili muscles, which both have important functions in the skin. The capillary networks drain into venules, small veins which carry the deoxygenated blood away from the skin and remove waste products.

The dermis is therefore well supplied with capillary blood vessels to bring nutrients and oxygen to the germinating cells in the epidermis and to remove waste products from them.

Lymph Vessels

The lymphatic vessels are numerous in the dermis and generally accompany the course of veins. They form a network through the dermis, allowing removal of waste from the skin's tissues. Lymph vessels are found around the dermal papillae, glands and hair follicle.

Nerves

Nerves are widely distributed throughout the dermis. Most nerves in the skin are sensory nerves, which send signals to the brain and are sensitive to heat, cold, pain, pressure, touch. Branched nerve endings, which lie in the papillary layer and hair root, respond to touch and temperature changes. Nerve endings in the dermal papillae are sensitive to gentle pressure and those in the reticular layer are responsive to deep pressure.

Sensory Nerves

There are at least five different types of sensory nerve endings in the skin.

- sensory
- touch
- pressure
- pain
- temperature

The sensory nerve endings are also called receptors because they are part of the nervous system at which information is received.

Touch Receptors

These receptors are located immediately below the epidermis. They are stimulated by light pressure on the skin which enables a person to distinguish between different textures such as rough, smooth, hard and soft.

Pressure Receptors

These receptors are situated beneath the dermis and stimulated by heavy pressure.

Pain Receptors

These receptors consist of branched nerve endings in the epidermis and dermis. They are quite evenly distributed throughout the skin and are important in that they provide a warning signal of damage or injury in the body.

Temperature Receptors

There are separate hot and cold receptors in the skin, that are stimulated by sudden changes in temperature.

The dermis also has motor nerve endings, which relay impulses from the brain and are responsible for the dilation and constriction of blood vessels, the secretion of perspiration from the sweat glands and the contraction of the arector pili muscles attached to hair follicles.

The Subcutaneous Layer

This is a thick layer of connective tissue found below the dermis. The tissues areolar and adipose are present in this layer to help support delicate structures such as blood vessels and nerve endings.

The subcutaneous layer contains the same collagen and elastin fibres as the dermis and also contains the major arteries and veins which supply the skin, forming a network throughout the dermis. The fat cells contained within this layer help to insulate the body by reducing heat loss. Below the subcutaneous layer of the skin lies the subdermal muscle layer.

KEY NOTE

As therapeutic treatments involve the stimulation of the tissues of the skin, they have an effect on both the superficial epidermis and the deeper dermis. One of the common skin reactions to treatments given is the creation of an **erythema**. This occurs when the blood capillaries just below the epidermis dilate and a reddening appears on the skin's surface. This indicates that the deeper layers of the skin such as the dermal and subdermal muscle layers are being stimulated in reaction to treatment.

TASK 1 – THE APPENDAGES OF THE SKIN

The skin has several appendages which are formed in the epidermis and extend down into the dermis. Complete the following table to identify the appendage of the skin and its functional significance.

Appendage	Structure	Location	Functional Significance
	small, smooth weak muscle made up of sensory fibres, which is associated with the hair	attached at an angle to the base of a hair follicle	
	dead keratinised structure which grows from a hair follicle and is made up of three layers	all over the body, with the exception of the palms of the hands and the soles of the feet	
	pore-like indentation in the epidermis	extends from the epidermis down to the dermis and surrounds the root of the hair	
	small sac-like pouch	majority open into a hair follicle but some open onto the skin surface. Found all over the body, except for the soles of the feet and the palms of the hands. They are more numerous on the scalp, face, chest and back	
	simple coiled tubular glands that open directly onto the surface of the skin	widely distributed over the surface of the skin; are most numerous in the palms of the hands and the soles of the feet	
	coiled tube that is connected with hair follicles	found only in the genital and underarm regions	

TASK 2 – THE STRUCTURE OF THE SKIN

Label the following parts on Figure 7:

epidermis	dermis	subcutaneous layer	eccrine sweat gland
hair	hair follicle	pore	erector pili muscle
sebaceous gland	adipose tissue	artery	vein
capillary network	motor nerve	pressure receptor	heat receptor
cold receptors	touch receptors	pain receptors	subdermal muscle layers

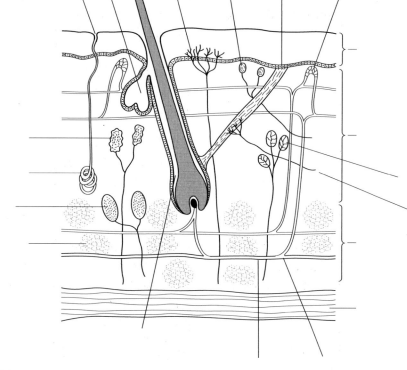

Figure 7
Cross section of the skin

Factors Affecting the Skin

There are many factors which affect the skin.

Diet

A healthy body is needed for a healthy skin. The skin can be thought of as a barometer of the body's general health. A nutritionally balanced diet is vital to the skin, as it will contain all the essential nutrients to nourish the cells in growth and repair.

In addition to the essential nutrients in the diet, the following vitamins aid the regeneration of the skin:

 Vitamin A: helps repair the body's tissues and helps prevent dryness and ageing
 Vitamin B: helps improve the circulation and the skin's colour, and is essential for cellular oxidation
 Vitamin C: is essential for healing and to maintain levels of collagen in the skin.

Water

Drinking an adequate amount of water (approximately six to eight glasses per day) aids the digestive system and helps to prevent a build up of toxicity in the tissues of the skin.

Sleep

Sleep is essential to physical and emotional well being and is one of the most effective regenerators for the skin.

Stress and Tension

When the body is subjected to regular stress and tension, it can cause sensitivity and allergies in the skin, as well as encourage the formation of lines around the eyes and the mouth.

Exercise

Regular exercise promotes good circulation, increased oxygen intake and blood flow to the skin.

Alcohol

Alcohol has a dehydrating effect on the skin, and excess consumption causes the blood vessels in the skin to dilate.

Smoking

Smoking affects the skin's cells, and destroys vitamins B and C which are important for a healthy skin. Smoking dulls the skin by polluting the pores, and increases the formation of lines around the eyes and the mouth.

Medication

Medication can affect the skin by causing dehydration, oedema, sensitivity and / or allergies. The action of ultra violet light, whether from a natural or artificial source can be hazardous to the skin.

- UVA penetrates deep into the dermis where it can cause premature ageing of the skin. Highly reactive molecules called free radicals are formed which cause skin cells to degenerate and lose their elasticity.
- UVB can cause sunburn, in which the cells become red and damaged and the skin may blister. UVB is also implicated in types of skin cancers, particularly malignant melanoma.

Chemicals

Harsh alkaline chemicals in products containing detergents and soaps can cause moisture loss in the skin, by stripping sebum from the skin's surface.

Climate

The climate has several effects on the skin:

- A cold climate can lead to a reduced production of sebum, resulting in reduced protection for the skin and increased water evaporation.
- A hot climate can lead to increased moisture loss through perspiration.
- Extremes of temperature (from hot to cold, or cold to hot) can lead to the formation of broken capillaries.

Environment

Air pollution such as carbon from smoke, fumes from car exhausts, chemicals from factories and the environment, can all be potentially damaging to the skin's cells. Environmental pollutants such as lead, mercury, aluminium can all accumulate in the body, and may find their way into food through polluted waters, rain and dust.

Hormones

The natural glandular changes of the body have an effect on the condition of the skin throughout life.

▶ During puberty, the sex hormones stimulate the sebaceous glands which may cause some imbalance in the skin.

▶ At the onset of menstruation, the skin may erupt due to the adjustment of hormone levels at that time.

▶ During pregnancy, pigmentation changes may occur, but usually disappear at birth.

▶ During the menopause, the activity of the sebaceous glands is reduced and the skin becomes drier.

Age

The natural process of ageing naturally affects the skin. From the mid-thirties, the skin starts to lose its firmness, and fine lines and wrinkles start to appear. In the forties and fifties, lines and wrinkles will deepen, and loss of muscle tone causes sagging of the skin on the cheeks and the neck. The connective tissue in the skin loses its elasticity and becomes less firm, and the skin becomes thinner and finer.

As part of the ageing process the process of cell regeneration in the skin decreases and the skin appears dry and dull.

The Functions of the Skin

The skin is so much more than an external covering. It is a highly sensitive boundary between our bodies and the environment. The skin has several important functions, in that it offers protection, temperature regulation, waste removal, as well as providing us with a sense of touch.

Protection

The skin acts as a protective organ in the following ways:

▶ the film of sebum and sweat on the surface of the skin, known as the acid mantle, acts as an anti-bacterial agent to help prevent the multiplication of micro-organisms on the skin

▶ the fat cells in the subcutaneous layer of the skin help protect bones and major organs from injury

▶ melanin, which is produced in the basal cell layer of the skin, helps to protect the body from the harmful effects of ultra violet radiation

▶ the cells in the horny layer of the skin overlap like scales to prevent micro-organisms from penetrating the skin and to prevent excessive water loss from the body.

Temperature Regulation

The skin helps to regulate body temperature in the following ways:

- when the body is losing too much heat, the blood capillaries near the skin surface contract, to keep warm blood away from the surface of the skin and closer to major organs
- the arector pili muscles raise the hairs and trap air next to the skin when heat needs to be retained
- the adipose tissue in the dermis and the subcutaneous layer helps to insulate the body against heat loss
- when the body is too warm, the blood capillaries dilate to allow warm blood to flow near to the surface of the skin, in order to cool the body
- the evaporation of sweat from the surface of the skin will also assist in cooling the body

Sensitivity

The skin is considered as an extension of the nervous system. It is very sensitive to various stimuli due to its many sensory nerve endings which can detect changes in temperature, pressure and register pain.

Excretion

The skin functions as a mini excretory system, eliminating waste through perspiration.

The eccrine glands of the skin produce sweat, which helps to remove some waste materials from the skin such as urea, uric acid, ammonia and lactic acid.

Storage

The skin also acts as a storage depot for fat and water. About 15 per cent of the body's fluids are stored in the subcutaneous layer.

Absorption

The skin has limited absorption properties. Substances which can be absorbed by the epidermis include fat soluble substances such as oxygen, carbon dioxide, fat soluble vitamins, steroids, along with small amounts of water.

KEY NOTE

The skin is capable of absorbing small particles of essential oils due to the fact that they contain fat and water soluble particles.

Vitamin D Production

The skin synthesises vitamin D when exposed to ultra-violet light. Modified cholesterol molecules in the skin are converted by the ultra violet rays in sunlight to vitamin D. It is then absorbed by the body for the maintenance of bones and the absorption of calcium and phosphorus in the diet.

Skin Diseases and Disorders

A therapist needs to be able to recognise diseases and disorders of the skin to enable proper judgement to be made as to whether a client may be treated in a particular area or whether they need to be referred to their medical practitioner for treatment and advice.

A therapist must therefore be particularly vigilant in recognising infectious skin diseases and disorders to prevent cross infection and to maintain a hygienic environment for themselves, their colleagues and clients.

Glossary of Useful Terms Associated with The Skin

Allergic Reaction

A disorder in which the body becomes hypersensitive to a particular allergen. When irritated by an allergen, the body produces histamine in the skin, as part of the body's defence or immune system.

The effects of different allergens are diverse and they affect different tissues and organs. For example, certain cosmetics and chemicals can cause rashes and irritation in the skin; certain allergens such as pollen, fur, feathers, mould and dust can cause asthma and hay fever. If severe, allergies may be extremely serious and result in anaphylactic shock.

Comedone

A collection of sebum, keratinised cells and wastes which accumulate in the entrance of a hair follicle. It may be open or closed. An open comedone is a 'blackhead' contained within the follicle, whereas a closed comedone is a whitehead, trapped underneath the skin's surface.

Cyst

An abnormal sac containing liquid or a semi-solid substance. Most cysts are harmless.

Erythema

Reddening of the skin due to the dilation of blood capillaries just below the epidermis in the dermis.

Fissure

A crack in the epidermis exposing the dermis.

Keloid

An overgrowth of an existing scar which grows much larger than the original wound. The surface may be smooth, shiny or ridged. The onset is gradual and is due to an accumulation or increase in collagen in the immediate area. The colour varies from red, fading to pink and white.

Lesion

A zone of tissue with impaired function, as a result of damage by disease or wounding.

Macule

A small flat patch of increased pigmentation or discolouration, for example, a freckle.

Milia

Sebum trapped in a blind duct with no surface opening. Usually found around the eye area. They appear as pearly, white hard nodules under the skin.

Mole

Moles are also known as a pigmented naevi. They appear as round, smooth lumps on the surface of the skin. They may be flat or raised and vary in size and colour from pink to brown or black. They may have hairs growing out of them.

Naevus

A mass of dilated capillaries. May be pigmented as in a birthmark.

Papule

Small raised elevation on the skin, less than 1cm in diameter, which may be red in colour. Often develops into a pustule.

Pustule

Small raised elevation on the skin which contains pus.

Skintag

Small growths of fibrous tissue, which stand up from the skin and sometimes are pigmented (black or brown).

Scar

A mark left on the skin after a wound has healed. Scars are formed from replacement tissue during the healing of a wound. Depending on the type and extent of damage, the scar may be raised (hypertrophic), rough and pitted (ice pick), or fibrous and lumpy (keloid). Scar tissue may appear smooth and shiny or form a depression in the surface.

Telangiecstasis

This is the term for dilated capillaries, where there is persistent vaso-dilation of capillaries in the skin. Usually caused by extremes of temperature and over-stimulation of the tissues, although sensitive and fair skins are more susceptible to this condition.

Tumour

A tumour is formed by an overgrowth of cells. Almost every type of cell in the epidermis and dermis is capable of benign or malignant overgrowth. Tumours are lumpy, and even when they cannot be seen, they can be felt underneath the surface of the skin.

Ulcer

A break or open sore in the skin extending to all its layers.

Vesicles

Small sac-like blisters. A bulla is a vesicle larger than 0.5cm and is commonly called a blister.

Wart

Well defined benign tumour which varies in size and shape. See Viral Infections page 45.

Weal

A raised area of skin, containing fluid which is white in the centre with a red edge. Is seen in the condition Urticaria (see page 49).

Disorders of the Sebaceous Gland

Acne Vulgaris

A common inflammatory disorder of the sebaceous glands which leads to the overproduction of sebum. It involves the face, back and chest and is characterised by the presence of comedones, papules, and in more severe cases, cysts and scars. Acne Vulgaris is primarily androgen-induced. It appears most frequently at puberty and usually persists for a considerable period of time.

Rosacea

A chronic inflammatory disease of the face in which the skin appears abnormally red. The condition is gradual, and begins with flushing of the cheeks and nose. As the condition progresses, it can become pustular.

Aggravating factors include hot, spicy foods, hot drinks, alcohol, menopause, the weather and stress.

Sebaceous Cyst

A round, nodular lesion with a smooth shiny surface, which develops from a sebaceous gland. They are usually found on the face, neck, scalp and back. They are situated in the dermis and vary in size from 5–50mm. The cause is unknown.

Seborrhoea

This condition is defined as an excessive secretion of sebum by the sebaceous glands. The glands are enlarged and the skin appears greasy, especially on the nose and the centre zone of the face.

The condition may develop into Acne Vulgaris and is common at puberty, lasting for a few years.

Disorders of the Sweat Glands

Hyperhidrosis

Excessive production of sweat affecting the hands, feet and underarms.

Bacterial Infections

Boil

A boil begins as a small inflamed nodule which forms a pocket of bacteria around the base of a hair follicle, or a break in the skin. Local injury or lowered constitutional resistance may encourage the development of boils.

Conjunctivitis

This is a bacterial infection following irritation of the conjunctiva of the eye. In this condition, the inner eyelid and eyeball appear red and sore, and there may be a pus-like discharge from the eye. The infection spreads by contact with the secretions from the eye of the infected person.

Folliculitis

This bacterial infection occurs in the hair follicles of the skin and appears as a small pustule at the base of a hair follicle. There is redness, swelling and pain around the hair follicle.

Impetigo

A superficial contagious inflammatory disease caused by streptococcal and staphylococcal bacteria. It is commonly seen on the face and around the ears, and features include weeping blisters which dry to form honey-coloured crusts. This bacteria is easily transmitted by dirty fingernails and towels.

Stye

Acute inflammation of a gland at the base of an eyelash, caused by bacterial infection. The gland becomes hard and tender, and a pus-filled cyst develops at the centre.

Viral Infections of the Skin

Herpes Simplex (cold sores)

Herpes Simplex is normally found on the face and around the lips. It begins as an itching sensation, followed by erythema and a group of small blisters; which then weep and form crusts.

This condition will generally persist for approximately two or three weeks, but will reappear at times of stress, ill health or exposure to sunlight.

Herpes Zoster (shingles)

Painful infection along the sensory nerves by the virus that causes chicken pox. Lesions resemble herpes simplex, with erythema and blisters along the lines of the nerves. Areas affected are mostly on the back or upper chest wall.

This condition is very painful due to acute inflammation of one or more of the peripheral nerves. Severe pain may persist at the site of shingles for months or even years after the apparent healing of the skin.

Warts

A wart is a benign growth on the skin caused by infection with the human papilloma virus.

- **Plane warts** are smooth in texture with a flat top, and are usually found on the face, forehead, back of the hands and the front of the knees.
- **Plantar warts** or **verrucae** occur on the soles of the feet and are usually the size of a pea.

Fungal Infections of the Skin

Ringworm

A fungal infection of the skin, which begins as small red papules that gradually increase in size to form a ring. Affected areas on the body vary in severity from mild scaling to inflamed itchy areas.

Tinea Capitis

This is a type of ringworm, and is a fungal infection of the scalp. It appears as painless, round, hairless patches on the scalp. Itching may be present and the lesion may appear red and scaly.

Tinea Pedis (athletes' foot)

This is a highly contagious condition which is easily transmitted in damp, moist conditions such as swimming pools, saunas and showers. Athletes foot appears as flaking skin between the toes which becomes soft and soggy. The skin may also split and the soles of the feet may occasionally be affected.

Infestation Disorders of the Skin

Pediculosis (lice)

This condition is commonly known as 'Lice' and is a contagious parasitic infection, where the lice live off the blood sucked from the skin. Head lice are frequently seen in young children and if not dealt with quickly, may lead to a secondary infection as a result of scratching (Impetigo). With head lice, nits may be found in the hair; they are pearl-grey or brown, oval structures found on the hair shaft close to the scalp. The scalp may appear red and raw due to scratching.

Body lice are rarely seen. They will occur on an individual with poor personal hygiene, and will live and reproduce in seams and fibres of clothing, feeding off the skin. Lesions may appear as papules, scabs and in severe cases pigmented dry, scaly skin. Secondary bacterial infection is often present.

A client affected by body lice will complain of itching, especially in the shoulder, back and buttock area.

Scabies

A contagious parasitic skin condition, caused by the female mite who burrows into the horny layer of the skin where she lays her eggs. The first noticeable symptoms of this condition is severe itching which worsens at night; papules, pustules and crusted lesions may also develop.

Common sites for this infestation are the ulnar borders of the hand, palms of the hands and between the fingers and toes. Other sites include the axillary folds, buttocks, breasts in the female and external genitalia in the male.

Pigmentation Disorders

Albinism

A condition in which there is an inherited absence of pigmentation in the skin, hair and eyes, resulting in white hair, pink skin and eyes. The pink colour is produced by underlying blood vessels which are normally masked by pigment. Other clinical signs of this condition include poor eyesight and sensitivity to light.

Chloasma

This is a pigmentation disorder which presents with irregular areas of increased pigmentation, usually on the face. It commonly occurs during pregnancy, and sometimes when taking the contraceptive pill due to stimulation of melanin by the female hormone oestrogen.

Lentigo

Also known as 'liver spots'. These are flat dark patches of pigmentation which are found mainly in the elderly, on skin exposed to light.

Vitiligo

This condition is presented on areas of the skin which lack pigmentation due to the basal cell layer of the epidermis no longer producing melanin. The cause of vitiligo is unknown.

Naevi

A naevi is a clearly defined malformation of the skin. There are many different types of naevi:

Portwine Stain

Also known as a 'deep capillary naevus'. Present at birth and may vary in colour from pale pink to deep purple. Has an irregular shape, but is not raised above the skin's surface. Usually found on the face, but may also appear on other areas of the body.

Spider Naevi

A collection of dilated capillaries which radiate from a central papule. Often appear during pregnancy or as the result from 'picking a spot'.

Strawberry Mark

Usually develops before or shortly after a baby is born, but disappears spontaneously before the child reaches the age of ten. It is raised above the skin's surface.

Hypertrophic Disorders

Hypertrophic skin disorders refer to conditions which have resulted in an increase of size of a tissue or organ. This is caused by an enlargement of the cells.

Malignant Melanoma

A malignant melanoma is a deeply pigmented mole which is life-threatening if it is not recognised and treated promptly. Its main characteristic is a blue-black module which increases in size, shape and colour, and is most commonly found on the head, neck and trunk.

Over-exposure to strong sunlight is a major cause, and its incidence is increased in young people with fair skins.

Rodent Ulcer

This is a malignant tumour, which starts off as a slow-growing pearly nodule, often at the site of a previous skin injury. As the nodule enlarges, the centre ulcerates and refuses to heal. The centre becomes depressed, and the rolled edges become translucent, revealing many tiny blood vessels. Rodent ulcers do not disappear and if left untreated may invade the underlying bone.

This is the most common form of skin cancer.

Squamous Cell Carcinoma

This is a malignant tumour which arises from the prickle cell layer of the epidermis. It is hard and warty and eventually develops a 'heaped-up, cauliflower' appearance. It is most frequently seen in elderly people.

Inflammatory Skin Conditions

Contact Dermatitis

Dermatitis literally means *inflammation* of the skin. 'Contact dermatitis' is caused by a primary irritant which causes the skin to become red, dry and inflamed. Substances which are likely to cause this reaction include acids, alkalis, solvent, perfumes, lanolin, detergent and nickels. There may be skin *infection* as well.

Eczema

A mild to chronic inflammatory skin condition characterised by itchiness, redness and the presence of small blisters that may be dry or weep, if the surface is scratched. Eczema is non contagious; cause may be genetic or due to internal and external influences.

It can cause scaly and thickened skin, mainly at flexures; for example, cubital area of the elbows and the back of the knees.

Psoriasis

A chronic inflammatory skin condition. Psoriasis may be recognised as the development of well-defined red plaques, varying in size and shape, and covered by white or silvery scales. Any area of the body may be affected by psoriasis, but the most commonly affected sites are the face, elbows, knees, nails, chest and abdomen. It can also affect the scalp, joints and nails.

Psoriais is aggravated by stress and trauma but is improved by exposure to sunlight.

Seborrhoeic Dermatitis

This is a mild to chronic inflammatory disease of hairy areas, well supplied with sebaceous glands. Common sites are the scalp, face, axilla, and in the groin. The skin may appear to have a grey tinge or may be dirty yellow in colour. Clinical signs include slight redness, scaling and dandruff in the eyebrows.

Urticaria

Also known as 'hives'. In this condition lesions appear rapidly and disappear within minutes or gradually over a number of hours. The clinical signs are the development of red weals which may later become white. The area becomes itchy or may sting.

There are a number of causes of urticaria, some of which are an allergic reaction to certain foods, for example, strawberries, shellfish, penicillin, house dust and pet fur. Other causes include stress and sensitivity to light, heat or cold.

TASK 3 – SKIN DISEASES AND DISORDERS

Complete the following table to identify the skin disease/disorder from the descriptions given and state the cause(s).

Disease/ Disorder	Description	Cause	Action
	inflammatory disorder of the sebaceous gland. Characterised by the presence of oily skin, comedones, papules, pustules, and in severe cases cysts and scars		Refer to GP. Condition may be treatable with GP supervision.
	irritation of the eye in which the eyelid and eyeball appears red; may also be a pus-like discharge from the eyes		Avoid contact. Refer client to GP.

Disease/ Disorder	Description	Cause	Action
	excessive secretion of sebum by the sebaceous glands. The glands are enlarged, the skin appears greasy especially in the nose and the centre zone of the face. The condition may develop into acne vulgaris and is common at puberty		GP referral may be advisable. Frequent cleansing required.
	chronic inflammatory disorder of the face in which the skin appears abnormally red		Refer to GP. Avoid stimulating treatments.
	inflamed nodule which forms a pocket of bacteria around the base of a hair follicle or a break in the skin		Avoid contact. Refer client to GP.
	inflammatory disease caused by streptococcal and staphylococcal bacteria. It is commonly seen on the face and around the ears and features include weeping blisters which dry to form honey-coloured crusts.		Avoid contact. Refer client to GP.
	normally found on the face and around the lips. It begins as an itching sensation, followed by erythema and a group of small blisters; they weep and form crusts.		Avoid contact. Refer client to GP.
	presents as flaking skin between the toes which becomes soft and soggy		Avoid contact. Refer client to GP.

Disease/ Disorder	Description	Cause	Action
	presents as small red papules that gradually increase in size to form a ring		Avoid contact. Refer client to GP.
	parasitic skin condition, caused by the female mite who burrows into the horny layer of the skin. Symptoms include severe itching; papules, pustules and crusted lesions may also develop		Avoid contact. Refer client to GP.
	areas of the skin which lack pigmentation due to the basal cell layer no longer producing melanin		No special precautions, other than advising client to avoid exposure of affected areas to UV light.
	pigmentation disorder which presents with areas of increased pigmentation, usually on the face		No special precautions, other than advising client to avoid exposure of affected areas to UV light.
	mild to chronic inflammatory skin condition characterised by itchy, red skin with the presence of small blisters that may be dry or weep if scratched		GP referral if chronic. Avoid affected area to avoid irritation. Care needed when selecting products.
	inflammation of the skin, which becomes red, dry and inflamed. Infection may also be present		GP referral if chronic. Avoid affected area to avoid irritation. Care needed when selecting products.

Disease/ Disorder	Description	Cause	Action
	chronic inflammatory skin condition, presents as well defined plaques covered by white or silvery scales. Commonly affected sites are the face, elbows, knees, nails, chest and the abdomen		GP referral if chronic. Avoid infected area to avoid irritation. Care needed when selecting products.
	benign growth on the skin; smooth in texture with a flat top. Usually found on the face, forehead, back of hands and front of knees		Avoid contact. Refer client to GP.
	benign growth on the skin, occurs on the soles of the feet and is usually the size of a pea		Avoid contact. Refer client to GP.

SELF ASSESSMENT QUESTIONS – THE SKIN

1.

a) List the five layers of the epidermis

b) Briefly describe the structure of each layer

c) In each case, state if the cells in that layer are living or dead

2. Briefly describe the process of cell regeneration in the epidermis, noting the functional significance of the relevant parts.

3. Name the protein found in the epidermis. What is its function?

4. What is melanin and where is it found in the skin?

5. Briefly state how the following factors affect the skin

a) age

b) hormones

c) diet

d) medication

6. Name the two layers of the dermis.

7. State the types of tissue the dermis contains.

8. What is the function of the dermis?

9. Name three types of cells present in the dermis.

10. Where is muscle located in relation to the layers of the skin?

11. List the sensations which the sensory nerves of the skin enable us to feel.

12. What is the function of the subcutaneous layer of the skin?

13. Briefly describe the primary functions of the skin.

CHAPTER 3

The Hair

In order to be successful in removing hair temporarily or permanently, it is imperative for a therapist to have a working knowledge of the anatomical structure of a hair.

By the end of this chapter you will be able to relate the following to your practical work carried out in the salon:

▶ the structural and functional parts of a hair
▶ the growth cycle of a hair
▶ the different types of hair growth on the body
▶ disorders of the hair.

Hair is an important appendage of the skin which grows from a sac-like depression in the epidermis called a hair follicle. Hair grows all over the body, with the exception of the palms of the hands and the soles of the feet and is a sexual characteristic.

The Function of Hair

The primary functions of hair are:

Physical Protection

▶ The eyelashes act as a line of defence by preventing the entry of foreign particles into the eyes and helping shade the eyes from the sun's rays.

▶ Eyebrow hairs help to divert water and other chemical substances away from the eyes.

▶ Hairs lining the ears and the nose trap dust and help to prevent bacteria from entering the body.

▶ Body hair acts as a protective barrier against the sun and helps to protect us against the cold with the help of the erector pili muscle.

Preventing Friction

▶ Body hair is present on the body where muscular action causes friction.

The Structure of a Hair

The hair is composed mainly of the protein keratin and therefore is a dead structure.

Longitudinally the hair is divided into three parts:

1 hair **shaft**: the part of the hair which lies above the surface of the skin
2 hair **root**: the part which is found below the surface of the skin
3 hair **bulb**: the enlarged part at the base of the hair root

Internally the hair has three layers, which all develop from the matrix (the active growing part of the hair):

▶ **cuticle**: the outer layer, made up of transparent protective scales which overlap one another. The cuticle protects the cortex and gives the hair its elasticity

▶ **cortex**: the middle layer, made up of tightly packed keratinised cells containing the pigment melanin which gives the hair its colour. The cortex helps to give strength to the hair

▶ **medulla**: the inner layer, made up of loosely connected keratinised cells and tiny air spaces. This layer of the hair determines the sheen and colour of hair due to the reflection of light through the air spaces

Hair Colour

Hair colour is due to the presence of melanin in the cortex and medulla of the hair shaft. In addition to the standard black colour, the melanocytes in the hair bulb produce two colour variations of melanin, brown and yellow.

▶ Blond, light coloured and red has a high proportion of the yellow variant.

▶ Brown and black hair possesses more of the brown and black melanin.

KEY NOTE

The hair turns grey when the melanocytes in the hair bulb stop producing melanin.

Individual Parts of Structure

The individual parts of a hair's structure are as follows:

Connective Tissue Sheath

The connective tissue sheath surrounds both the follicle and sebaceous gland. Its function is to supply the follicle with nerves and blood. It is the main source of sustenance for the follicle.

Outer Root Sheath

The outer root sheath forms the follicle wall and is continuous with the basal cell layer of the epidermis. It provides a permanent source of growing cells (hair germ cells) to enable the follicle to grow and renew cells during its life cycle.

Dermal Papilla

The dermal papilla is an elevation at the base of the bulb, which contains a rich blood supply. It serves as a crucial source of nourishment for hair, providing the hair cells with food and oxygen.

Inner Root Sheath

The inner root sheath originates from the dermal papilla at the base of the follicle and grows upwards with the hair. The inner root sheath is made up of similar cells to the cuticle of the hair; they lie in the opposite direction facing downwards towards the dermal papilla. This allows the cells of the inner and outer layer to lock together, thereby shaping and contouring the hair, helping to anchor it into the follicle. The inner root sheath ceases to grow when level with the sebaceous gland.

Hair Bulb

The hair bulb is the enlarged part at the base of the hair root. A gap at the base leads to a cavity which contains the dermal papilla. The hair bulb is where the cells grow and divide by the process of mitosis.

Matrix

The matrix is located at the lower part of the hair bulb. It is the area of mitotic activity of the hair cells.

TASK 1 – THE STRUCTURE OF A HAIR

Complete the following table to identify the structure and its functional significance of the individual parts of a hair structure

Structure	Location	Functional Significance
	elevation at the base of the bulb	
	originates from the base of the follicle and grows up with the hair	
	forms the follicle wall and is continuous with the basal cell layer of the epidermis	
	surrounds follicle and sebaceous gland	
	lower part of the hair bulb	
	enlarged area at the base of the hair root	

TASK 2 – THE HAIR IN ITS FOLLICLE

Label the following parts on Figure 8:

hair shaft   erector pili muscle   outer root sheath

matrix   connective tissue sheath   hair bulb   dermal papilla

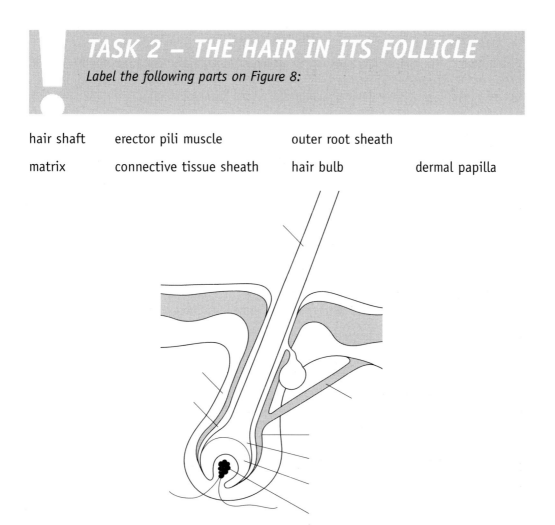

Figure 8
A hair in its follicle

Types of Hair

There are three main types of hair in the body: lanugo, vellus and terminal hair.

Lanugo Hair

Lanugo is the fine, soft hair found on a foetus. It grows from around the third to the fifth month of pregnancy and is eventually shed to be replaced by secondary vellus hairs, around the seventh to eight month of the pregnancy.

Lanugo hair is often unpigmented and as it is soft and fine, lacks a medulla.

Vellus Hair

Vellus hair is the soft downy hair found all over the face and body, except for the palms of the hands, soles of the feet, eyelids and lips. Vellus hairs are often unpigmented and do not have a medulla or a well developed bulb.

Vellus hair lies close to the surface of the skin and therefore has a shallow follicle. If stimulated by an increase in blood circulation resulting from hormonal changes in the body (such as puberty, pregnancy or menopause) or medication, the shallow follicle of a vellus hair can grow downwards to become a coarse, dark terminal hair.

Terminal Hair

Terminal hairs are the longer, coarser hairs found on the scalp, under the arms, eyebrows, pubic regions, arms and legs; most are pigmented. They vary greatly in shape, diameter, length, colour and texture.

They are deeply seated in the dermis and have well-defined bulbs.

KEY NOTE

When terminal hairs such as in the bikini line or underarm are removed from their follicles, there may be the incidence of minor blood spotting, because these hairs grow from deep follicles with a strong blood supply. It is therefore extremely important to carry out all the relevant hygienic precautions, as the risk of infection is increased in these areas. It is also very important to ensure client comfort by stretching the skin, as removal of these hairs from deep rooted follicles can cause discomfort.

Factors About Hair Growth

▶ Hair begins to form in the foetus from the third month of pregnancy.
▶ Growth of hair originates from the matrix, which is the active growing area where cells divide and reproduce by mitosis.
▶ Living cells, which are produced in the matrix, are pushed upwards away from their source of nutrition, they die and are converted to keratin to produce a hair.
▶ Hair has a growth pattern which ranges from approximately four to five months for an eyelash hair to approximately four to seven years for a scalp hair.
▶ Hair growth is affected by illness, diet and hormonal influences.

The Growth Cycle of a Hair

Each hair has its own growth cycle and undergoes three distinct stages of development.

Anagen

▶ active growing stage
▶ lasts from a few months to several years
▶ hair germ cells reproduce at matrix
▶ new follicle is produced which extends in depth and width
▶ the hair cells pass upwards to form hair bulb
▶ hair cells continue rising up the follicle and as they pass through the bulb they differentiate to form individual structures of hair
▶ inner root sheath grows up with the hair, anchoring it into the follicle
▶ when cells reach upper part of bulb they become keratinised
▶ two-thirds of way up follicle, hair leaves inner root sheath and emerges onto surface of skin

KEY NOTE

An anagen hair receives its nourishment to grow from dermal papilla and when removed has a visibly developed bulb and inner root sheath intact.

Catagen

▶ lasts approximately two to four weeks
▶ transitional stage – from active to resting
▶ hair separates from dermal papilla and moves slowly up the follicle
▶ follicle below retreating hair shrinks
▶ hair rises to just below level of sebaceous gland where the inner root sheath dissolves and the hair can be brushed out

KEY NOTE

In a catagen hair, a column of epithelial cells remains in contact with the dermal papilla. As the hair breaks away from the bulb, it receives its nourishment from the follicle wall. A catagen will have no bulb visible when removed and will appear shorter and dehydrated.

Telogen

▶ short resting stage
▶ shortened follicle rests until stimulated again
▶ hair is shed onto skin's surface
▶ new replacement hair begins to grow

KEY NOTE

A telogen hair has a diminished blood supply and when removed has small brim-like fibres at its end.

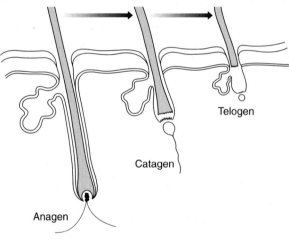

Figure 9
The hair growth cycle

KEY NOTE

Each follicle has a life expectancy of its own. Once a follicle has produced its expected number of hairs, it will no longer function.

Different Types of Hair Growth

As there is a continuous cycle of hair growth, the amount of hair on the body remains fairly constant. However, hair growth will vary from client to client and from area to area. A new client coming to the salon for a hair removal treatment should be made aware of the fact that hair growth occurs in three stages, which will result in the hair being at different lengths both above the skin and below it.

KEY NOTE – INGROWING HAIRS

Ingrowing hairs present a problem when carrying out hair removal treatments as the hair grows back on itself when the follicle is blocked. This may be because the hair has grown so weak that it can no longer push up through the skin, so it grows parallel to the skin.

Skin Reaction to Hair Removal Treatments

While carrying out hair removal treatments, it is important to remember that the hair follicle is part of the skin's structure, therefore any treatment which affects the hair is also going to affect the skin.

Once a hair has been removed, the maximum amount of blood will be sent straight to the area being treated to heal and protect the skin. This is a normal reaction of the skin, and extra blood that has been sent to the treated area will soon be diverted again within a few hours of treatment. As the treated area of skin will have open follicles, it is vital that a client adheres strictly to after-care advice specified as open follicles offer bacteria an easy entry into the body.

Disorders of the Hair

▶ **Folliculitis** is a bacterial infection of the hair follicles of the skin and appears as a small pustule at the base of a hair follicle. There is redness, swelling and pain around the hair follicle.
▶ **Pediculosis (Lice)** is an infestation (wingless insect) that feeds on human and animal blood. With head lice, nits may be found in the hair. Nits are pearl-grey or brown, oval structures found on the hair shaft close to the scalp. The scalp may appear red and raw due to scratching.
  A client, affected by body lice will complain of itching especially in the shoulder, back and buttock area.
▶ **Tinea Capitis** is a fungal infection of the scalp (Ringworm) and appears as painless, round, hairless patches on the scalp. Itching may be present and the lesion may appear red and scaly.

SELF ASSESSMENT QUESTIONS – THE HAIR

1. Name the three layers of the hair.

2. Which layer/s of the hair:

a) contain melanin

b) contain keratinised cells

c) is the outer most layer?

3. Where are the hair cells produced?

4. Where does hair growth originate from?

5. State three factors that affect hair growth.

6. Which structure supplies the hair follicle with nerves and blood?

7. Which structure provides a crucial source of nourishment for the hair?

8. Which structure grows upwards with the hair, and what is its function?

    ....

    ....

9. Name two appendages of the skin associated with the hair. State their key function in relation to the hair.

    ....

    ....

    ....

    ....

10. State the three stages of hair growth and briefly state their significance.

    ....

    ....

    ....

11. How would you recognise a growing hair after it has been removed?

    ....

    ....

12. Describe the skin's reaction to hair removal treatments.

    ....

    ....

    ....

CHAPTER 4

The Nail

A sound knowledge of the structure of the nail and its functional parts is fundamental to understanding how a nail grows.

A competent therapist needs to be able to:

▶ understand the process of nail growth to explain the benefits of salon treatments and the consequences any damage may have
▶ correctly identify the functional parts of the nail to apply treatments correctly and avoid any resulting damage.

By the end of this chapter, you will be able to relate the following knowledge to your practical work carried out in the salon:

▶ the structure of the nail in its bed
▶ the functional significance of the individual parts of the nail
▶ the process of nail growth
▶ adverse treatable nail conditions and non-treatable nail diseases.

The nail is an important appendage of the skin and is an extension of the clear layer of the epidermis. It is composed of horny flattened cells which undergo a process of keratinisation, giving the nail a hard appearance.

It is the protein keratin which helps to make the nail a strong but flexible structure. The part of the nail the eye can see is dead as it has no direct supply of blood, lymph and nerves; all nutrients are supplied to the nail via the dermis.

Functions of the Nail

The nail has several important functions:

▶ It forms a protective covering at the ends of the phalangeal joints of the fingers and the toes, helping to support the delicate network of blood vessels and nerves at the end of the fingers.

▶ It is a useful tool in that it enables us to concentrate touch to manipulate small objects and to scratch surfaces.

The Structure of the Nail

The nail has several important anatomical regions:

Matrix

The matrix is situated immediately below the cuticle and is the nail's most important feature. It is the area where the living cells are produced.

The matrix receives a rich supply of blood which supplies oxygen to the nail, and is vital to the production of new cells. It is the area from which the health of the nail is determined.

KEY NOTE

If the matrix is deprived of nutrients, the nail may have impaired growth or be malformed.

Nail Bed

The nail bed is situated immediately below the nail plate and is a continuation of the matrix. It is the part of the skin upon which the nail plate rests.

The nail bed is richly supplied with blood vessels, lymph vessels and nerves from the underlying dermis. The key functions of the nail bed are to provide nourishment and protection for the nail.

KEY NOTE

The nail bed gives a healthy nail its pink colour.

Cuticle

The cuticle is a fold of overlapping skin that surrounds the base of the nail. There are different names given to the different areas of the cuticle:

- the **eponychium** is the dead cuticle that adheres to the base of the nail, near the lanula
- the **peronychium** is the cuticle that outlines the nail plate
- the **hyponychium** is the cuticle skin found under the free edge of the nail.

The key function of the cuticle is to protect the matrix and provide a protective seal against bacteria.

KEY NOTE

The cuticle is an extension of the horny layer of the epidermis.

Lanula

The lanula is the light-coloured semicircular area of the nail, commonly called the half moon, that lies in between the matrix and the nail plate. The lanula is always present but not always visible, as it may be obscured by the cuticle.

It is an area of the nail where cells start to harden; the cells here are in a transitional stage (between hard and soft). The lanula is therefore a bridge between the living cells of the matrix and the dead cells of the nail plate.

KEY NOTE

The lanula is always present but not always visible, as it may be obscured by the cuticle.

Nail Plate

The nail plate is the main visible part of the nail which rests on the nail bed and ends at the free edge. It is made up of layers of translucent dead, keratinised cells to make the nail hard and strong. The layers of cells are packed very closely together, with fat but very little moisture.

The key function of the nail plate is to offer protection for the nail bed.

> ### KEY NOTE
>
> There are no blood vessels or nerves in the nail plate, which explains why a nail may be cut without bleeding or pain.

Nail Walls

The nail walls are the folds of skin overlapping the sides of the nails. They surround three sides of the nail and are firmly attached to the sides of the nail plate.

The key function of the nail walls is to protect the edges of the nail plate from external damage.

> ### KEY NOTE
>
> The nail walls help to hold the nail in place and give it stability.

Nail Grooves

The nail grooves are the deep ridges under the sides of the nail. As the nail grows along with the nail bed, it passes along the nail grooves, which guide it and help it to grow straight.

> ### KEY NOTE
>
> The nail grooves help to keep the nail growing forward in a straight line.

Free Edge

The free edge is the part of the nail plate that extends beyond the nail bed. It is the part of the nail that is filed and usually the hardest part.

> ### KEY NOTE
>
> As the free edge is not in contact with the underlying tissue, it lacks the colour of the rest of the nail plate.

TASK 1 – THE STRUCTURE OF THE NAIL

Complete the following table to identify the part of the nail and its functional significance.

Part of Nail	Location	Functional Significance
	immediately below the cuticle	
	surrounds the base of the nail; is an extension of the horny layer of the epidermis	
	deep ridges under the sides of the nail	
	immediately below the nail plate, continuation of the matrix	
	between the matrix and the nail plate	
	skin attached to the sides of the nail plate	
	main visible part of the nail which rests on the nail bed and ends at the free edge.	

TASK 2 – CROSS SECTION OF A NAIL

Label the following parts on Figure 10:

matrix   nail bed   cuticle   lanula

nail plate   nail wall   nail grooves   free edge

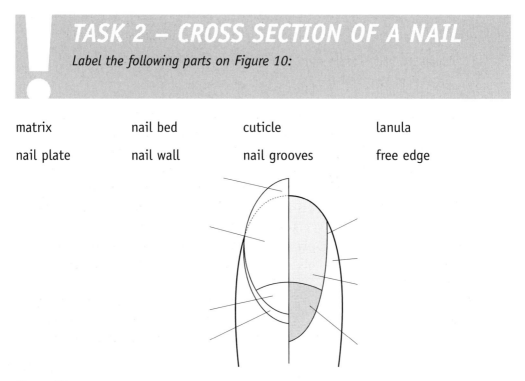

Figure 10
Cross section of a nail

Factors about Nail Growth

▶ Nails start growing on a foetus before the fourth month of pregnancy.

▶ Nail growth occurs from the nail matrix by cell division.

▶ As new cells are produced in the matrix, older cells are pushed forward and are hardened by the process of keratinisation, which forms the hardened nail plate.

▶ As the nail grows it moves along the nail grooves at the sides of the nail, which helps to direct the nail growth along the nail bed.

▶ It takes approximately six months for cells to travel from the lanula to the free edge of the nail.

▶ The growth of a nail does not follow a growth cycle and hence growth is continuous throughout life.

▶ The average growth rate of a nail is approximately 3mm per month.

▶ The growth rate of nails will vary from person to person and from finger to finger, with the index finger generally being the fastest to grow.

▶ Toe nails have a slower rate of growth than fingernails.

▶ The rate of growth of a nail is faster in the summer due to an increase in cell division as a result of exposure to ultra violet radiation.

▶ A good blood supply is essential to nail growth; oxygen and nutrients are fed to the living cells of the nail matrix and nail bed.

▶ Protein and calcium are good sources of nourishment for the nails.

Nail growth may be affected by the following factors:

▶ **ill health**: during illness, your body will receive a reduced blood supply to the nails as it attempts to restore the rest of the body to good health.

▶ **diet**: a nutritional deficiency can result in a diminished blood supply to the nail.

▶ **age**: during ageing, the growth of a nail slows down due to the fact that the blood vessels supplying the matrix and the nail become less efficient.

▶ **poor technique**: if a heavy pressure is used when using manicure implements such as a cuticle knife, damage may be caused to the matrix cells resulting in ridges to the nail. This may only be temporary, as new cells produced in the matrix will replace the damaged ones, and depending on the extent of the damage the ridges may eventually grow out.

▶ **an accident**: such as shutting a finger in the door. This may result in bruising and bleeding of the nail or even the complete removal of a nail. It could result in permanent malformation of the nail if the nail bed has become damaged.

KEY NOTE

It is essential when carrying out treatments for the hands and feet that care is taken to avoid damaging the nail matrix and nail bed, as they are the fundamental parts of the nail concerned with growth.

Nail Diseases and Nail Disorders

Diseases of the nail are as a direct result of bacteria, fungi, parasites or viruses attacking the nail or surrounding tissues.

Nail **disorders** may be caused by illness, physical and chemical damage, by general neglect or by poor manicuring techniques. Nail disorders do not contra-indicate manicure or pedicure treatments, however, nail diseases do, as they may cause cross infection.

A therapist must be able to recognise diseases and disorders in order that the correct treatment or advice may be given.

Common Nail Diseases

Paronychia

Inflammation of the skin surrounding the nail. The tissues may be swollen; pus may be present which can develop into an abscess. It is a common condition on the fingers, and is caused by bacterial or viral infection.

- Initially the cause may be due to prolonged immersion of the hands in water, poor manicure techniques, picking the cuticle or the nail wall separating from the nail.
- Infection with the herpes simplex virus can give rise to a whitlow, an abscess that forms around the nail.

Onychomycosis

This is a term given to fungal infections of the nail, commonly called Ringworm. It attacks the nail bed and nail plate, and presents as white or yellow scaly deposits at the free edge, which may spread down to invade the nail walls or bed. The nails become thickened, brittle, opaque or discoloured as a result of this condition. The nail plate will appear spongy and furrowed.

- In its advanced stages, the nail plate may separate from the nail bed (a condition known as Onycholysis (see below).
- There may also be accompanying dryness and skin scaling at the base of the fingers and on the palms.

Onychia

This is a generic term used to describe any disease of the nail, but more specifically refers to inflammation of the nail bed. In this condition, the nail matrix appears red. There may be swelling, tenderness and pus formation. This could lead to the nail being shed.

- This condition may be caused by wearing false nails for too long, or by harsh manicuring, chemical applications, or by a variety of infections or physical damage.

Onycholysis

This is separation or loosening of part or all of a nail from its bed. It may be due to disease, physical damage, or may occur spontaneously without any apparent cause. It can occur if sharp instruments are used under the free edge. Penetration of the flesh line allows bacteria or other infection to enter the nail bed.

Nail Disorders

Leuconychia

This is a term given to white or colourless nails, or nails with white spots, streaks or bands. There may also be evidence of ridging. It may be caused as a result of injury to the matrix or the effects of disease.

▶ The white spots will usually disappear as the nail grows.

Onychophagy

This is the technical term for nail biting in which the free edge, nail plate and cuticle are bitten to leave the hyponychium exposed and the cuticle and surrounding skin ragged, inflamed and sore.

▶ Nail biting is usually a nervous or stress-induced habit.

Nail Ridges / Corrugations

Ridges in the nail may occur due to irregular formation of the nail or to physical / chemical injury of the nail matrix.

▶ Ridges may be vertical, which are common in healthy nails due to uneven development of the nail tissue, poor manicuring techniques or the effects of harsh chemicals.
▶ Ridges may also be horizontal and can be indicative of abnormal nail growth, a symptom of body malfunction or disease.
▶ Deep horizontal lines are often associated with illness.

Hang Nail

A hang nail is a small strip of skin that hangs loosely at the side of the nail, or a small portion of the nail itself splitting away. A hang nail may develop due to dry, torn or split cuticles.

▶ Common causes are hands being immersed in water for long periods, cutting the nails too close, digging the cuticles, improper filing or the effects of detergents and other chemicals.

Onychogryphosis

This is a term given to an ingrown fingernail or toe nail. The first signs are inflammation, followed by tenderness, swelling and pain. Infection may aggravate the condition.

▶ This condition is caused by ill-fitting shoes, cutting or filing nails too short or too close to the skin. It may also be due to a malformation of the nail when it was beginning to grow.

Pterygium

This is a condition where the cuticle becomes overgrown and excessive, and grows forward. The cuticle at the base of the nail becomes dry and split and grows forward sticking to the nail plate.

▶ Pterygium may be due to faulty nail care or lack of nail care.

Koilonychia

This is the term given to concave spoon-shaped nails. In this condition the nails are thin, soft and hollowed. Koilonychia may be congenital, or it may be due to lack of iron or other minerals.

▶ The spoon-shape results from abnormal growth at the nail matrix.

Eggshell Nails

This is a term given to thin, white nails that are more flexible than normal. In this condition the nail separates from the nail bed and curves at the free edges. The condition may be associated with illness.

Onychorrhexis

This is the term given to dry, brittle nails. In this condition the nail loses its moisture, becomes dry and the free edge splits. The nails may peel into layers very easily. There may be transverse or longitudinal splitting of the nail plate, and inflammation, tenderness, pain, swelling and infection may be present.

▶ Frequent immersion in water and contact with detergents and chemicals contribute to this condition. It may also indicate an iron deficiency, anaemia, or incorrect filing which causes the nail plate to split.

TASK 3 – COMMON NAIL DISEASES

A therapist must be able to recognise diseases and disorders in order that the correct treatment or advice may be given. Identify the following nail diseases from the description. State the action to be taken in each case.

Disease	Description	Cause	Action
	inflammation of the skin surrounding the nail. Tissues appear swollen; pus may be present.	bacterial or viral infection. Contributory factors include: prolonged immersion of hands in water, poor manicure techniques, picking the cuticle.	
	nail matrix appears red; there may be swelling, tenderness and pus formation. Could lead to the nail being shed.	Wearing false nails for too long, harsh manicuring, chemical applications or by a variety of infections or physical damage.	
	presents as white, or yellow scaly deposits at the free edge. Condition may spread to the nail walls or bed. Nails become thickened, brittle, opaque or discoloured and nail plate will appear spongy and furrowed.	fungal infection.	

SELF ASSESSMENT QUESTIONS – THE NAIL

1. What are the functions of the nail?

2. What substance is the nail mainly composed of?

3. The nail is an extension of which layer of the epidermis?

4. Which layer of the skin supplies nutrients to the nail?

5. Identify the following from the description

a) the part that helps direct the nail along the nail bed

b) the part that determines the health of the nail

c) the part that supplies nourishment and protection for the nail

d) the area where cells start to keratinise or harden

e) the part that protects the matrix and provides a protective seal against bacteria

f) the part that holds the nail in place, protecting the edges of the nail plate from external damage

6. Briefly describe the process of nail growth.

CHAPTER 5

The Skeletal System

The skeleton is made up of no fewer than 206 individual bones, which collectively form a strong framework for the body. Bones are like *landmarks* in the body, and by tracing their outlines you can be accurate in describing the position of muscles, glands and organs in relation to your work as a therapist. Learning the positions of the bones of the skeleton is essential for learning the position of the superficial muscles, as bones must have muscle attachments to enable them to move.

On page 100 you will find a glossary of anatomical terms to assist you in learning the precise positions of bones.

A competent therapist needs to be able to:

▶ have a good knowledge of the framework of the body in order to be able to apply treatments safely and effectively

By the end of this chapter you will be able to relate the following knowledge to your practical work carried out in the salon:

▶ the functions of the skeleton
▶ the names and positions of the primary bones of the skeleton
▶ anatomical terms used to describe the position of the bones of the skeleton
▶ the composition and types of bone tissue
▶ the classifications of the different types of bones
▶ the growth and development of bone

▶ the importance of good posture
▶ disorders of the skeletal system.

Composition of Bone

Bone Tissue

Bone is one of the hardest types of connective tissue in the body, and when fully developed is composed of water, calcium salts and organic matter. Bone tissue is a type of living tissue that is made from special cells called **osteoblasts**. There are two main types of bone tissue: *compact* and *cancellous*.

All bones have both types of tissue, the amount being dependent on the type of bone.

Compact (dense) Bone

This is the hard portion of the bone that makes up the main shaft of the long bones and the outer layer of other bones. It protects spongy bone and provides a firm framework for the bone and the body. The bone cells in this type of bone (osteocytes) are located in concentric rings around a **central haversian canal**, through which nerves, blood and lymphatic vessels pass.

Cancellous (spongy) Bone

In contrast, this is lighter in weight than compact bone. It has an open sponge-like appearance, and is found at the ends of long bones or at the centre of other bones.

It does not have a haversian system, but consists of a web-like arrangement of spaces that are filled with red bone marrow and separated by the thin processes of bone. Blood vessels run through every layer of cancellous bone, conveying nutrients and oxygen.

Bone Marrow

Bones contain two types of marrow: red and yellow.

▶ Red marrow manufactures red blood cells. It is found at the end of long bones and at the centre of other bones of the thorax and pelvis.
▶ Yellow marrow is found chiefly in the central cavities of long bones.

Except for the ends that form joints, bones are covered with a thin membrane of connective tissue called the **periosteum**. The outer layer of the periosteum is extremely dense and contains a large number of blood vessels. The inner layer contains osteoblasts and fewer blood vessels. The periosteum provides attachment for muscles, tendons and ligaments.

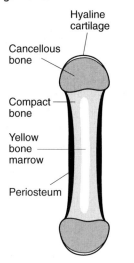

Figure 11
The structure of a long bone

Ossification of Bone

The process of bone development is called **ossification**. This process begins in the embryo near the end of the second month, and is not complete until about the 25th year of life.

Ossification takes place in three stages:

▶ the cartilage-forming cells (called chondrocytes) enlarge and arrange themselves in rows similar to the bone they will eventually form.
▶ Calcium salts are then laid down by special bone building cells called osteoblasts.
▶ A second set of cells called osteoclasts (known as cartilage-destroying cells) bring about an antagonistic action, enabling the absorption of any unwanted bone.

A fine balance of osteoblast and osteoclast activity helps to maintain the formation of normal bone. Osteocytes are mature bone cells that maintain bone during our lifetime.

Types of Bone

Bones are classified according to their shape. The classifications are long bones, short bones, flat bones, irregular bones and sesamoid bones.

Long Bones

Long bones have a long shaft (a diaphysis) and one or more endings, or swellings (epiphysis). Smooth hyaline cartilage covers the articular surfaces of the shaft endings. Between the diaphysis and epiphysis of growing bone is a flat plate of hyaline cartilage called the epiphyseal cartilage or **growth plate**. This is the site of bone growth, and as fast as this cartilage grows it is turned into bone, allowing the bone to continue to grow in length.

▶ A growth spurt is often seen during puberty, through the influence of the sex hormones oestrogen and testosterone, both of which promote the growth of long bone.
▶ At around 25 years of age, the entire plate becomes ossified.

KEY NOTE

Children's bones are more flexible as their bodies contain more cartilage and soft bone cells, since complete calcification has not yet taken place. In older adults this process is reversed, as bone cells outnumber cartilage cells, and bone becomes more brittle due to the fact it contains more minerals and fewer blood vessels. This explains why elderly people's bones are more prone to fracture and slower to heal.

All bones of the limbs are long bones (except the wrist and ankle bones).

Short Bones

These bones are generally cube shaped, with their lengths and widths being roughly equal. The bones of the wrist and the ankle are examples of short bones.

Flat Bones

Flat bones are platelike structures with broad surfaces. Examples include the ribs and the scapulae.

Irregular Bones

Irregular bones have a variety of shapes. Examples include the vertebrae and some of the facial bones.

Sesamoid Bones

These are small rounded bones that are embedded in a tendon. The largest sesamoid bone is the patella, which is embedded in the quadriceps femoris tendon.

Functions of the Skeleton

The skeletal system is made up of all types of bones which form the skeleton or bony framework of the body. Before learning the individual bones of the skeleton, it is important to understand the functions of the skeleton as a whole.

Support

The skeleton bears the weight of all other tissues; without it we would be unable to stand up. Consider the bones of the vertebral column, the pelvis, the feet and the legs which all support the weight of the body.

Shape

The bones of the skeleton give shape to structures such as the skull, thorax and limbs; without soft tissue covering these areas we would look prehistoric!

Protection of Vital Organs and Delicate Tissue

The skeleton surrounds vital organs and tissue with a tough, resilient covering. For example, the rib cage protects the heart and lungs and the vertebral column protects the spinal cord.

Attachments for Muscles and Tendons

To allow movements. Bones are like anchors which allow the muscle to function efficiently.

Movement

This happens as a result of co-ordinated action of muscles upon bones and joints. Bones are therefore levers for muscles.

Formation of Blood Cells

These develop in red bone marrow.

Mineral Reservoir

The skeleton acts as a storage depot for important minerals such as calcium, which can be released when needed for essential metabolic processes such as muscle contraction and conduction of nerve impulses.

The skeletal system is divided into two parts. The **Axial Skeleton** forms the main axis or central core of the body and consists of the following parts:

▶ The skull
▶ The vertebral column
▶ The sternum
▶ The ribs.

The **Appendicular Skeleton** supports the appendages or limbs and gives them attachment to the rest of the body. It consists of the following parts:

▶ The shoulder girdle
▶ Bones of the upper limbs
▶ Bones of the lower limbs
▶ Bones of the pelvic girdle.

The Skull

The skull rests upon the upper end of the vertebral column and weighs around 11 pounds!

It consists of 22 bones:

▶ eight bones that make up the skull or cranium
▶ 13 forming the facial skeleton.

The skull encloses and protects the brain and provides a surface attachment for various muscles of the skull. The eight bones of the skull are as follows:

▶ One **Frontal*** bone forms the anterior part of the roof of the skull, the forehead and the upper part of the orbits or eye sockets. Within the frontal bones are the two frontal sinuses, one above each eye near the midline.
▶ Two **Parietal*** bones form the upper sides of the skull and the back of the roof of the skull.
▶ Two **Temporal*** bones form the sides of the skull below the parietal bones and above and around the ears. The temporal bone contributes to part of the cheekbone via the zygomatic arch (formed by the zygomatic and temporal bones). Located behind the ear and below the line of the temporal bones are the mastoid processes, to which the stermomastoid muscles of the neck are attached.
▶ One **Sphenoid** bone is located in front of the temporal bone and serves as a bridge between the cranium and the facial bones. It articulates with the frontal, temporal, occipital and ethmoid bones.
▶ One **Ethmoid** bone forms part of the wall of the orbit, the roof of the nasal cavity and part of the nasal septum.
▶ The **Occipital*** bone forms the back of the skull.

The bones marked with a * are considered to be the primary bones of the skull

> ### KEY NOTE
>
> There are many openings present in the bones of the skull which act as passages for blood vessels and nerves entering and leaving the cranial cavity. For example, there is a large opening at the base of the skull called the **foramen magnum** through which the spinal cord and blood vessels pass to and from the brain.

The Bones of the Face

There are 14 facial bones in total. These are mainly in pairs, one on either side of the face:

- **Two Maxillae*** – these are the largest bones of the face and they form the upper jaw and support the upper teeth. The two maxillary bones fuse to become one early in life. An important part of the maxillae are the maxillary sinuses, which open into the nasal cavity.

- **One Mandible*** – this is the only moveable bone of the skull and forms the lower jaw and supports the lower teeth. The mandible is the largest and heaviest bone in the skull.
- **Two Zygomatic*** – these are the most prominent of the facial bones and they form the cheekbones.
- **Two Nasal** – these small bones form the bridge of the nose.
- **Two Lacrimal** – these are the smallest of the facial bones, and are located close to the medial part of the orbital cavity.
- **Two Turbinate** – these are layers of bone located either side of the outer walls of the nasal cavities.
- **One Vomer** – this is a single bone at the back of the nasal septum.
- **Two Palatine** – these are L-shaped bones which form the anterior part of the roof of the mouth.

The bones marked with a * are considered to be the primary bones of the face.

The Sinuses

There are four pairs of air-containing spaces in the skull and face called the sinuses. The function of the sinuses is to lighten the head, provide mucus and act as a resonance chamber for sound. The pairs of sinuses are named according to the facial bones by which they are located. They are the **frontal** sinuses, the **sphenoidal** sinuses, the **ethomoidal** sinuses, and the **maxillary** sinuses (which are the largest).

TASK 1 – MAJOR BONES OF THE SKULL AND FACE

Label the major bones of the skull and face on Figure 12:

temporal   maxilla   occipital   mandible   frontal   parietal   zygomatic

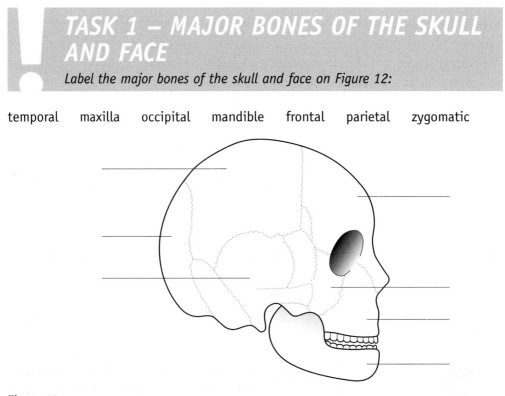

Figure 12
Major bones of the skull and face

The Vertebral Column

The vertebral column lies on the posterior of the skeleton, extending from the skull to the pelvis, providing a central axis to the body. It consists of 33 individual irregular bones called vertebrae; however, the bones of the base of the vertebral column, the sacrum and coccyx, are fused to give 24 movable bones in all.

The vertebral column is made up of the following:

▶ Seven cervical vertebrae

These are the vertebrae of the neck; the top two vertebrae, C1 the atlas, and C2 the axis are part of a pivot joint, which allows the head and neck to move freely. These are the smallest vertebrae in the vertebral column.

▶ Twelve thoracic vertebrae

These are the vertebrae of the mid spine and they lie in the thorax, where they articulate with the ribs. These vertebrae lie flatter and downwards to allow for muscular attachment of the large muscle groups of the back. They can be easily felt as you run your fingers down the spine.

- Five lumbar vertebrae

These lie in the lower back and are much larger in size than the vertebrae above them as they are designed to support more body weight. These vertebrae can be easily felt on the lower back due to their large shape and width.

- Five sacral vertebrae

This is a very flat triangular shaped bone and lies in between the pelvic bones. It is made up of five bones which are fused together. A characteristic feature of the sacrum is the sacral holes, of which there are eight. It is through these holes that nerves and blood vessels penetrate.

- Four coccygeal vertebrae

These are made up of four bones which are fused together and are sometimes referred to as the tail bone.

KEY NOTE

In between the vertebrae lies a padding of fibrocartilage called the intervertebral discs. These give the vertebrae a certain degree of flexibility and also act as shock absorbers in between the vertebrae, cushioning any mechanical stress that may be placed upon them.

Functions of the Vertebral Column

Now we have considered the individual structure of the vertebrae, let us consider the functions of the vertebral column as a whole:

- The vertebral column provides a strong and slightly flexible axis to the skeleton.
- By way of its different shaped vertebrae with their roughened surfaces, it is able to provide a surface for the attachment of muscle groups.
- The vertebral column also has a protective function as it protects the delicate nerve pathways of the spinal cord.

Therefore, the vertebral column mirrors the primary functions of the skeleton in its supportive and protective roles.

The Thorax

This is the area of the body enclosed by the ribs, which provides protection for the heart and lungs. Essential organs contained within this cavity include:

- the sternum
- the ribs
- 12 thoracic vertebrae.

The Sternum

This is commonly referred to as the breast bone and is a flat bone lying just beneath the skin in the centre of the chest. The sternum is divided into three parts:

- the manubrium, the top section
- the main body, the middle section
- the xiphoid process, the bottom section.

The top section of the sternum articulates with the clavicle and the first rib. The middle section articulates with the costal cartilages which link the ribs to the sternum. The bottom section provides a point of attachment for the muscles of the diaphragm and the abdominal wall.

The Ribs

There are 12 pairs of ribs. They articulate with the thoracic vertebrae, posteriorly. Anteriorly, the first ten pairs attach to the sternum via the costal cartilages, the first seven directly (known as the true ribs), the remaining three indirectly (known as the false ribs). The last two ribs have no anterior attachment and are called the false ribs.

The Appendicular Skeleton

This consists of the shoulder girdle, the bones of the upper and lower limbs and the pelvic girdle.

The Shoulder Girdle

The shoulder girdle connects the upper limbs with the thorax and consists of four bones.

- two scapula
- two clavicle.

The **scapula** is a large flat bone, triangular in outline, which forms the posterior part of the shoulder girdle. It is located between the second and the seventh rib.

The scapula articulates with the clavicle and the humerus and serves as a point of muscle attachment which connects the shoulder girdle with the trunk and upper limbs.

The **clavicle** is a long slender bone with a double curve. It forms the anterior portion of the shoulder girdle. At its medial end it articulates with the top part of the sternum and at its lateral end it articulates with the scapula. The clavicle acts as a brace to hold the arm away from the top of the thorax.

The clavicle provides the only bony link between the shoulder girdle and the axial skeleton. The arrangement of bones and the muscle attached to the scapula and the clavicle allow for a considerable amount of movement of the shoulder and the upper limbs.

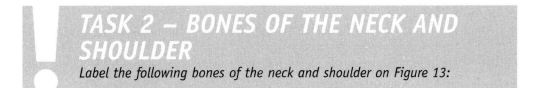

TASK 2 – BONES OF THE NECK AND SHOULDER

Label the following bones of the neck and shoulder on Figure 13:

clavicle   scapula   cervical vertebrae

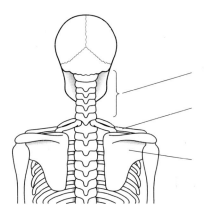

Figure 13
Bones of the neck and shoulder

The Upper Limb

The **upper limb** consists of the following bones:

▶ Humerus
▶ Radius
▶ Ulna

▶ Carpals
▶ Metacarpals
▶ Phalanges.

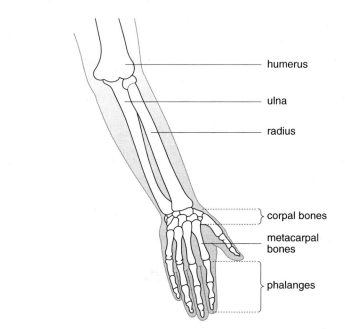

humerus

ulna

radius

corpal bones

metacarpal bones

phalanges

Figure 14
Bones of the upper limb

The **humerus** is the long bone of the upper arm. The head of the humerus articulates with the scapula, forming the shoulder joint. The distal end of the bone articulates with the radius and ulna to form the elbow joint.

The **ulna** and **radius** are the long bones of the forearm. The two bones are bound together by a fibrous ring, which allows a rotating movement in which the bones pass over each other. The ulna is the bone of the little finger side and is the longer of the two forearm bones. The radius is situated on the thumb side of the forearm and is shorter than the ulna. The joint between the ulna and the radius permits a movement called pronation. This is when the radius moves obliquely across the ulna so that the thumb side of the hand is closest to the body.

The movement called supination takes the thumb side of the hand to the lateral side. The radius and the ulna articulate with the humerus at the elbow and the carpal bones at the wrist.

The Wrist and Hand

The wrist consists of eight small bones of irregular size which are collectively called **carpals.** They fit closely together and are held in place by ligaments. The carpals are arranged in two groups of four; those of the upper row articulate with the ulna and the radius and the lower row articulates with the metacarpals. The upper row nearest the forearm is called scaphoid, lunate, triquetral and pisiform; the lower row is called the trapezium, trapezoid, capitate and hamate.

There are five long **metacarpal** bones in the palm of the hand; their proximal ends articulate with the wrist bones and the distal ends articulate with the finger bones. There are fourteen **phalanges**, which are the finger bones, two of which are in the thumb or pollex and three in each of the other digits.

The Lower Limb

The lower limb consists of the following bones:

- Femur
- Tibia
- Fibula
- Patella
- Tarsals
- Metatarsals
- Phalanges.

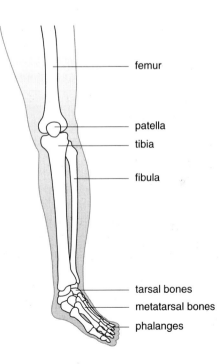

Figure 15
Bones of the lower limb

The **femur** is the longest bone in the body and has a shaft and two swellings at each end. The proximal swelling has a rounded head like a ball which fits into the socket of the pelvis to form the hip joint. Below the neck are swellings called trochanters which are sites for muscle attachment. The distal ends of the femur articulate with the patella or knee cap.

The Patella

The patella or kneecap is located anterior to the knee joint. Its main function is to provide stabilisation, cushion the hinge joint at the knee, and protect the knee by shielding it from impact.

The **tibia** and **fibula** are long bones of the lower leg. The tibia is situated on the anterior and medial side of the lower leg. It has a large head, where it joins the knee joint and the shaft leads down where it forms part of the ankle. The tibia is the larger of the two bones of the lower leg and thus carries the weight of the body. The fibula is situated on the lateral side of the tibia in the lower leg and is the shorter and thinner of the two bones. The end of the fibula forms part of the ankle on the lateral side.

The Foot

There are seven bones in the foot which are collectively called the tarsals. Each tarsal is an irregular bone that slides minutely over the next bone to collectively provide motion. The individual tarsals are as follows:

Talus

The talus bone is the main tarsal. It articulates with the tibia and fibula to form the ankle joint. The talus is significant in that it bears the weight of the entire body when standing or walking.

Calcaneum

The calcaneum is also known as the heel bone. It is the largest and most posterior tarsal bone. The calcaneum is an important site for attachment of muscles of the calf.

Cuboid

The cuboid is situated between the fourth and fifth metatarsals and the calcanuem on the lateral (outer) border of the foot.

Cuneiform

There are three cuneiform bones, which are located between the navicular bone and the first three metatarsal bones. They are numbered medially to laterally from I through to III (the most medial being I, the middle being II and the most lateral being III).

Navicular

The navicular bone is situated between the talus bone and the three cuneiforms. There are five metatarsals forming the dorsal surface of the foot. Fourteen phalanges form the toes, two of which are in the hallux or big toe and three to each of the other digits.

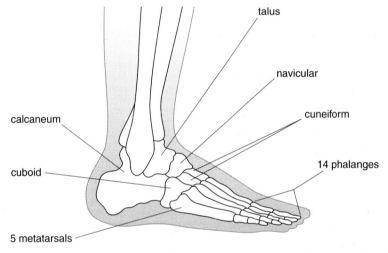

Figure 16
Bones of the foot

Arches of the Foot

The bones of the feet form arches which are designed to support body weight and to provide leverage when walking.

The arches of the foot are maintained by ligaments and muscles. They give the foot resilience in bearing the body's weight when running or walking.

There are four arches of the foot.

1 **Medial longitudinal arch**, which runs along the medial side of the foot from the calcaneum bone to the end of the metatarsals.
2 **Lateral longitudinal arch**, which runs along the lateral side of the foot from the calcaneum bone to the end of the metatarsals.
3 **Anterior transverse arch**, which runs across the top of the foot, across the lower end of the metatarsals.
4 **Posterior transverse arch**, which runs across the bottom of the foot, across the lower end of the metatarsals.

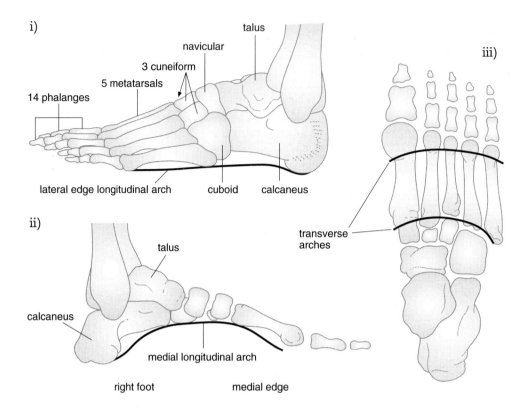

Figure 17
*Arches of the foot (i) lateral
longitudinal arch (ii) medial
longitudinal arch (iii) transverse arch*

The Pelvic Girdle

The pelvic girdle consists of two hip bones which are joined together at the back by the sacrum and at the front by the symphysis pubis. Each hip bone consists of three separate ones which are fused together:

▶ the ilium
▶ the ischium
▶ the pubis.

The ilium is the largest and most superior pelvic bone in the pelvic girdle. Its upper border is called the iliac crest, which is an important site of attachment for muscles of the anterior and posterior abdominal walls.

The ischium is the bone at the lower end and forms the posterior part of the pelvic girdle.

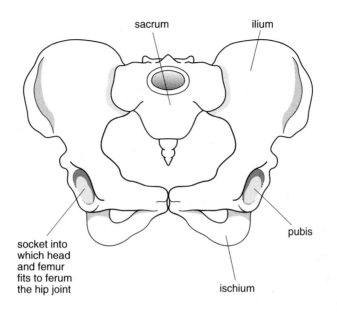

Figure 18
Bones of the pelvic girdle

KEY NOTE

The **ischial tuberosity** is a bony protrusion which is the part of the ischium that you sit on. It receives the weight of the body when sitting, and provides muscle attachments for the muscles such as the hamstrings and the adductors. Poor sitting posture can result from sitting on your sacrum instead of the ischial tuberosity.

The pubis is the collective name for the two pubic bones in the most anterior portion of the pelvis. The two pubic bones resemble a wishbone and are linked via a piece of cartilage called the symphysis pubis. The pubic bones provide attachment sites for some of the abdominal muscles and fascia.

Functions of the Pelvic Girdle

Like the vertebral column, the pelvic girdle mirrors the primary functions of the skeleton, insofar as it has a role in *supporting* the vertebral column and the body's weight and that it offers *protection* by encasing delicate organs such as the uterus and bladder.

TASK 3 – PRIMARY BONES OF THE SKELETON

Label the following bones on Figures 19 and 20:

clavicle	sternum	ribs	ulna	radius	humerus
patella	femur	tibia	fibula	ilium	ischium
pubis	tarsals	metatarsals	phalanges	scapula	carpals
metacarpals	cervical vertebrae	thoracic vertebrae	lumbar vertebrae		coccyx
					sacrum

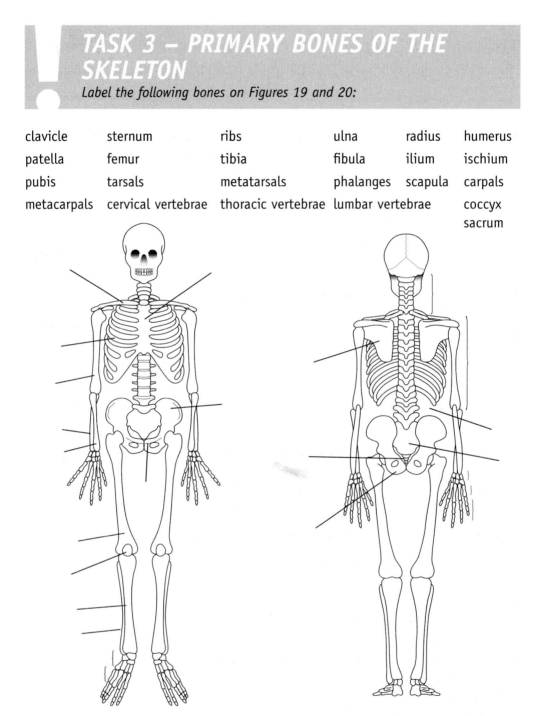

Figure 19

Anterior view of the primary bones of the skeleton

Figure 20

Posterior view of primary bones of the skeleton

Posture

Posture is a measure of balance and body alignment, and is the maintenance of strength and tone of the body's muscles against gravity. Good posture is said to be when the maximum efficiency of the body is maintained with the minimum effort.

When evaluating posture, an imaginary line is drawn vertically through the body – this is called the centre of gravity line. From the front or back this line should divide the body into two symmetrical halves:

▶ with feet together the ankles and knees should touch
▶ the hips should be the same height
▶ the shoulders should be level
▶ the sternum and vertebral column should run down the centre of the body in line with the centre of gravity line
▶ the head should be erect and not tilted to one side.

Posture varies considerably in individuals and is influenced by factors such as body frame size, heredity, occupation, habits and personality. Additional factors which may also affect posture include clothing, shoes and furniture.

Good posture is important as it:

▶ allows a full range of movement
▶ improves physical appearance
▶ keeps muscle action to a minimum, thereby conserving energy and reducing fatigue
▶ reduces the susceptibility of injuries
▶ aids the body's systems to function efficiently.

Poor posture may have the following effects on the body:

▶ produce alterations in body function and movement
▶ waste energy
▶ increase fatigue
▶ increase the risk of backache and headaches
▶ impair breathing
▶ increase the risk of muscular, ligament or joint injury
▶ affect circulation
▶ affect digestion
▶ give a poor physical appearance.

> ## KEY NOTE
>
> Poor posture or misalignment of the body is frequently found to be the cause of continued or chronic pain, as the body makes compensatory changes which are habit forming.

Disorders of the Skeletal System
Ankylosing Spondylitis

A systemic joint disease characterised by inflammation of the intervertebral disc spaces, costo-vertebral and sacroiliac joints. Fibrosis, calcification, ossification and stiffening of joints are common and the spine becomes rigid. Typically, a client will complain of persistent or intermittent lower back pain. Kyphosis is present when the thoracic or cervical regions of the spine are affected and the weight of the head compresses the vertebrae and bends the spine forward.

▶ This condition can cause muscular atrophy and loss of balance and falls.
▶ Typically this disease affects young male adults.

Arthritis – Gout

A joint disorder due to deposition of excessive uric acid crystals accummulating in the joint cavity.

▶ It commonly affects the peripheral joints, commonly the metatarsophalangeal joint of the big toe.
▶ Kidneys can be affected. Other cartilage may be involved including the ear pinna.

Arthritis – Osteoarthritis

A joint disease characterised by the breakdown of articular cartilage, growth of bony spikes, swelling of the surrounding synovial membrane, and stiffness and tenderness of the joint. Is also known as degenerative arthritis.

▶ It is common in the elderly and takes a progressive course.
▶ This condition involves varying degrees of joint pain, stiffness, limitation of movement, joint instability and deformity.
▶ It commonly affects the weight bearing joints – the hips, knees , the lumbar and cervical vertebrae.

Arthritis – Rheumatoid

Chronic inflammation of peripheral joints resulting in pain, stiffness and potential damage to joints. It can cause severe disability. Joint swellings and rheumatoid nodules are tender.

Bursitis

Inflammation of a bursa (small sac of fibrous tissue that is lined with synovial membrane and filled with synovial fluid). It usually results from injury or infection and produces pain, stiffness and tenderness of joint adjacent to the bursa.

Osteoporosis

Brittle bones due to ageing and the lack of hormone oestrogen which affects the ability to deposit calcium in the matrix of bone. This can also result from prolonged use of steroids. Vulnerability to osteoporosis can be inherited. Bones can break easily and vertebrae can collapse.

Temporo-mandibular Joint Tension (TMJ syndrome)

A collection of symptoms and signs produced by disorders of the temporo-mandibular joint. It is characterised by bilateral or unilateral muscle tenderness and reduced motion. It presents with a dull aching pain around the joint often radiating to the ear, face, neck or shoulder. The condition may start off as clicking sounds in the joint. There may be protrusion of the jaw or hypermobility and pain on opening the jaw. It slowly progresses to decreased mobility of the jaw and locking of the jaw may occur.

▶ Causes include chewing gum, biting nails, biting off large chunks of food, habitual protrusion of the jaw, tension in the muscles of the neck and back and clenching of the jaw. It may also be caused by injury and trauma to the joint or through a whiplash injury.

Postural Defects

Lordosis

This is an abnormally increased inward curvature of the lumbar spine. In this condition the pelvis tilts forward, and as the back is hollow, the abdomen and buttocks protrude, and the knees may be hyperextended.

▶ Typical problems associated with this condition are tightening of the back muscles followed by a weakening of the abdominal muscles. Because of the anterior tilt of the pelvis, hamstring problems are common.

▶ Increased weight gain or pregnancy may cause or exacerbate this condition.

Kyphosis

This is an abnormally increased outward curvature of the thoracic spine. In this condition the back appears round as the shoulders point forward and the head moves forward. A tightening of the pectoral muscles is common in this condition.

Scoliosis

This is a lateral curvature of the vertebral column, which may be to the left or right side. Evident signs of this condition include unequal leg length, distortion of the rib cage, unequal position of the hips or shoulders and curvature of the spine (usually in the thoracic region).

Glossary of Anatomical Terms

An anatomical position is determined from a central imaginary line running down the centre of the body. In order to give precise descriptions of the positions of certain bones and muscles, it is important to use anatomical terminology. Using anatomical terms is like learning a new language. It will enable you to state precisely where a bone or muscle is situated in the body:

Anterior: front surface of the body

Posterior: back surface of the body

Lateral: away from the midline

Medial: towards the midline

Superior: upper surface of a structure, towards the head

Inferior: below or lower surface of a structure, away from the head

Proximal: nearest to the midline or point of origin of a part

Distal: furthest away from the midline or point of origin of a part

Dorsal: furthest away from the midline or point of origin of a part

Dorsal: top of the foot is the dorsal surface

Plantar: sole of the foot is the plantar surface

SELF ASSESSMENT QUESTIONS – THE SKELETAL SYSTEM

1. List the functions of the skeleton.

2. List the two types of bone tissue in the skeleton.

3. Name the major bones of the skull.

4. Name the major bones of the face.

5. Name the bones of the shoulder girdle, stating their position.

6. List the different types of vertebrae in the spine, stating their position and number in each case.

7. What are the functions of the vertebral column?

8. Which bones are contained within the thorax?

9. Name the bones of the upper limb.

10. Name the bones of the lower limb.

11. State the structure and functions of the arches of the foot.

12. State the position and number of bones in the pelvic girdle.

13. What is the function of a ligament in the skeletal system?

....

....

....

CHAPTER 6

Joints

We have so far seen that the skeleton comprises bones which provide support and protection for our body. Bones must, however, be *linked* together in order to facilitate their supportive role and to allow movement to occur. It is joints that provide the link between bones of the skeletal system.

Joints therefore serve two purposes:

1 They hold bones together via ligaments offering stability for the joint
2 They afford the skeletal system more flexibility by facilitating movement, which is assisted by associated muscles and tendons

On page 112–116, you will find a glossary of terms to assist you in learning the angular movements that occur at joints.

A competent therapist needs to be able to:

▶ have a good knowledge of joints to understand how body movements occur

By the end of this chapter you will be able to relate the following knowledge to your practical work carried out in the salon:

▶ the type of joints and their range of movement
▶ the structure of a synovial joint
▶ anatomical terms used to describe the range of movements.

A joint is formed where two or more bones or cartilage meet and is otherwise known as an articulation. Where a bone is a *lever* in a movement, the joint is the **fulcrum** or the support which steadies the movement and allows the bone to move in certain directions.

Types of Joint

Joints are classified according to the *degree* of movement possible at each one. There are three main joint classifications:

1 Fibrous, where no movement is possible (also known as fixed joints)
2 Cartilaginous, where slight movement is possible
3 Synovial, which are freely moveable joints.

Fibrous Joints

These are immovable joints, which have tough fibrous tissue between the bones. Often the edges of the bones are dovetailed together into one another, as in the sutures of the skull. Some examples of fibrous joints include the joints between the teeth and between the maxilla and mandible of the jaw.

Cartilaginous Joints

These are slightly movable joints, which have a pad of fibrocartilage between the end of the bones making the joint. The pad acts as a shock absorber. Some examples of cartilaginous joints are those between the vertebrae of the spine, and at the symphysis pubis, in between the pubis bones.

Synovial Joints

These are freely movable joints which have a more complex structure than the fibrous or cartilaginous joints. Before looking at the different types of synovial joints it is important to have an understanding of the general structure of a synovial joint.

The General Structure of a Synovial Joint

- A synovial joint has a space between the articulating bones which is known as the synovial cavity.
- The surface of the articulating bones is covered by hyaline cartilage, which is supportive to the joint by providing a hard-wearing surface for the bones to move against one another with the minimum of friction.
- The synovial cavity and the cartilage are encased within a fibrous capsule, which helps to hold the bones together to enclose the joint. This joint capsule is

reinforced by tough sheets of connective tissue called ligaments, which bind the articular ends of bones together.

▶ The joint capsule is reinforced enough to allow strength to resist dislocation but is flexible enough to allow movement at the joint.

▶ The inner layer of the joint capsule is formed by the synovial membrane which secretes a sticky oily fluid called synovial fluid which lubricates the joint and nourishes the hyaline cartilage.

▶ As the hyaline cartilage does not have a direct blood supply, it relies on the synovial fluid to deliver its oxygen and nutrients and to remove waste from the joint, which is achieved via the synovial membrane.

TASK 1 – THE STRUCTURE OF A SYNOVIAL JOINT

Label the following parts of a synovial joint on Figure 21:

synovial cavity   hyaline cartilage   synovial membrane   bone

fibrous capsule

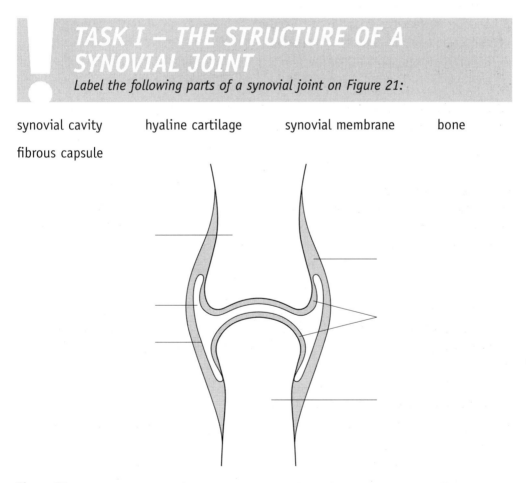

Figure 21
Structure of a synovial joint

Types of Synovial Joints

Synovial joints are classified into six different types according to their shape and the movements possible at each one. The degree of movement possible at each synovial joint is dependent on the type of synovial joint and its articulations.

Ball and Socket Joint

A ball and socket joint allows movement in many directions around a central point. This type of joint is formed when the rounded head of one bone fits into a cup-shaped cavity of another bone.

▶ Angular movements possible at this joint include flexion, extension, adduction, abduction, rotation and circumduction.

▶ Ball and socket joints allow the greatest freedom of movement, but are also the most easily dislocated. Examples includes the hip and the shoulder joints.

Hinge Joint

This type of joint is where the rounded surface of one bone fits the hollow surface of another bone. Movement is only possible in one direction.

▶ A hinge joint allows flexion and extension, changing the angle of the bones at the joint, as in a door hinge.

▶ Examples include the knee and the elbow joints, and the joints in between the phalanges.

Condyloid Joint

The joint surfaces of a condyloid joint are shaped so that the concave surface of one bone can slide over the convex surface of another bone in two directions. Although a condyloid joint allows movement in two directions, one movement dominates.

▶ Movements possible at a condyloid joint include flexion, extension, adduction and abduction.

▶ Examples include the wrist joint, and the joint between the metacarpals and phalanges (metacarpophlangeal joints).

Gliding Joint

Gliding joints permit gliding between two or more bones. They are often referred to as synovial plane joints, as they occur where two flat surfaces of bone slide against one another.

▌ Gliding joints allow only a gliding motion in various planes.

▌ Examples include the joints between the vertebrae and the sacroliliac joint.

Pivot Joint

A pivot joint occurs where a process of bone rotates in a socket (one component is shaped like a ring and the other component is shaped so that it can rotate within the ring).

▌ A pivot joint only permits rotation.

▌ Examples include the joint between the first and second cervical vertebrae (the atlas and the axis) and the joint at the proximal ends of the radius and the ulna.

Saddle Joint

This type of joint is shaped like a saddle. Its articulating surfaces of bone have both rounded and hollow surfaces so that the surface of one bone fits the complementary surface of the other.

▌ Movements possible at this joint include flexion, extension, adduction, abduction and a small degree of axial rotation.

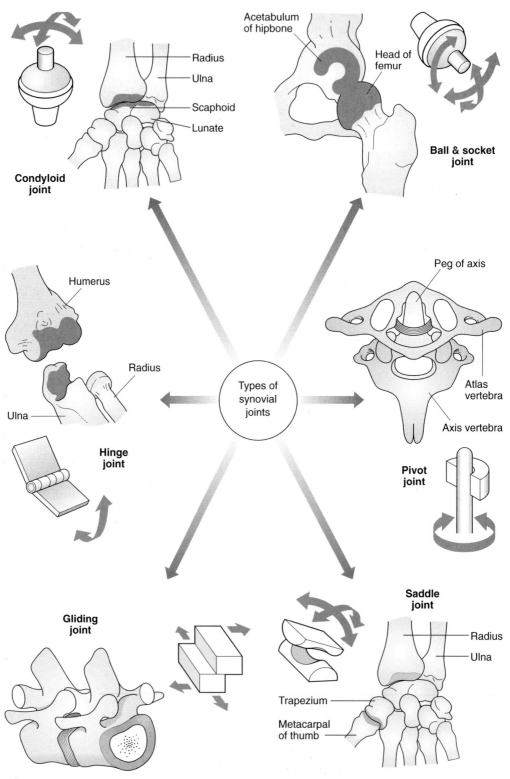

Figure 22

Types of synovial joints

Joint Disorders

See Disorders of the Skeletal System on pages 99–100.

> # ! TASK 2 – TYPES OF SYNOVIAL JOINT
>
> *Complete the following table to identify the type of joint and its location in the body from the descriptions given.*

Type of Joint	Description	Location	Movements Possible
	This type of joint is formed when the rounded head of one bone fits into a cup shaped cavity of another bone.		flexion, extension, adduction, abduction, rotation and circumduction
	This type of joint is where the rounded surface of one bone fits the hollow surface of another bone. Movement is only possible in one direction.		flexion and extension
	This type of joint is identified when the articulating surfaces of bone have both rounded and hollow surfaces so that the surface of one bone fits the complementary surface of the other.		flexion, extension, adduction, abduction and a small degree of axial rotation
	This joint occurs where a process of bone rotates in a socket.		rotation only
	The joint surfaces of this joint are shaped so that the concave surface of one bone can slide over the convex surface of another bone in two directions		flexion, extension, adduction and abduction

Glossary of Angular Movements Possible at Joints

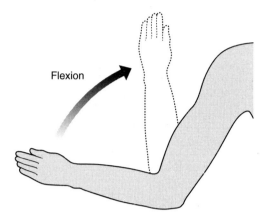

Figure 23

Flexion: bending of a body part at a joint so that the angle between the bones is decreased

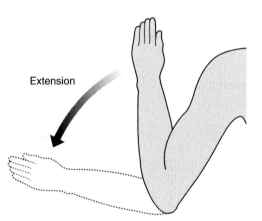

Figure 24

Extension: straightening of a body part at a joint so that the angle between the bones is increased

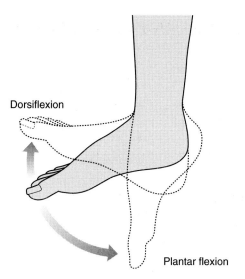

Figure 25

Dorsiflexion: upward movement of the foot so that feet point upwards
Plantar flexion: downward movement of the foot so that feet face downwards towards the ground

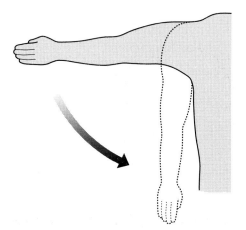

Figure 26

Adduction: movement of a limb towards the midline

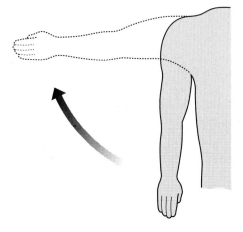

Figure 27

Abduction: movement of a limb away from the midline

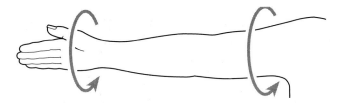

Figure 28

Rotation: movement of a bone around an axis (180 degrees)

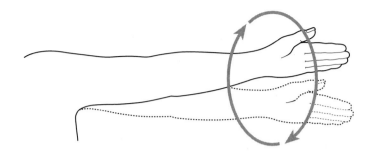

Figure 29

Circumduction: a circular movement of a joint (360 degrees)

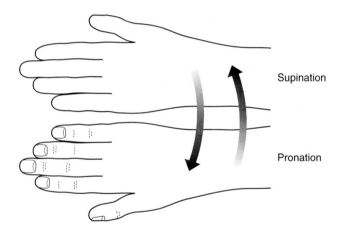

Figure 30

Supination: turning the hand so that the palm is facing upwards
Pronation: turning the hand so that the palm is facing downwards

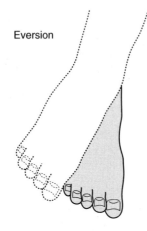

Eversion

Figure 31

Eversion: soles of the feet face outwards

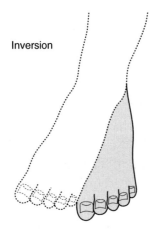

Inversion

Figure 32

Inversion: soles of the feet face inwards

? SELF ASSESSMENT QUESTIONS

1. Name the three main joint classifications

2. State the joint classifications for the following:
   (e.g. the hip is classified as a synovial joint but is also a ball-and-socket joint.)

a) joints between the skull bones

b) the atlas and axis

c) the symphysis pubis

d) the wrist joint

e) the knee joint

f) the trapezium and the metacarpal of the thumb

3. Describe the following terms

a) flexion

b) abduction

c) rotation

d) extension

e) supination

CHAPTER 7

The Muscular System

The muscular system comprises over 600 individual specialised cells called muscles which are primarily concerned with movement and body co-ordination. There is an intimate relationship between muscle and bone as both contribute to creating a movement in the body.

You will have learnt from the skeletal system that bones and joints provide the leverage in a movement, but it is in fact the muscle which provides the pull upon the bone to effect the movement. The key to learning the anatomical position and action of muscles is to first learn the individual position of the bones. It is then a logical step to learn the muscle attachments in relation to bone and what movements those muscles create. This chapter concentrates largely on the superficial muscles of the face and body as these are the muscles which cover the body and therefore are the ones upon which a therapist will be primarily working.

A competent therapist needs to be able to:

▶ have a good knowledge of the superficial muscles of the face and body in order to be able to carry out treatments safely and effectively

By the end of this chapter, you will be able to recall and understand the following in relation to your practical work:

▶ the functions of muscle
▶ the structure of voluntary muscle tissue

▶ muscle contraction
▶ muscle fatigue
▶ the effects of temperature and increased circulation on muscle contraction
▶ muscle tone
▶ disorders of the muscular system
▶ the anatomical position and action of the main superficial muscles of the face and body.

The Functions of Muscle

The muscular system consists largely of skeletal muscle tissue which covers the bones on the outside, and connective tissue which attaches muscles to the bones of the skeleton. Muscles, along with connective tissue, help to give the body its contoured shape.

The muscular system has three main functions:

Movement

Consider the action of picking up a pen that has dropped onto the floor. This seemingly simple action of retrieving the pen involves the co-ordinated action of several muscles pulling on bones at joints to create movement. Muscles are also involved in the movement of body fluids such as blood, lymph and urine. Consider also the beating of the heart which is continuous throughout life.

Maintaining Posture

Some fibres in a muscle resist movement and create slight tension in order to enable us to stand upright. This is essential, as without body posture we would be unable to maintain normal body positions such as sitting down or standing up.

The Production of Heat

As muscles create movement in the body they generate heat as a by product, which helps to maintain our normal body temperature.

Muscle Tissue

Muscle tissue makes up about 50 per cent of your total body weight and is composed of:

▶ 20 per cent protein
▶ 75 per cent water
▶ 5 per cent mineral salts, glycogen and fat.

There are three types of muscle tissue in the body:

1 **Skeletal** or voluntary muscle tissue which is primarily attached to bone
2 **Cardiac** muscle tissue which is found in the walls of the heart
3 **Smooth** or involuntary muscle tissue which is found inside the digestive and urinary tracts, as well as in the walls of blood vessels.

All three types of muscle tissue differ in their structure and functions and the degree of control the nervous system has upon them.

A therapist is primarily concerned with the structure of voluntary muscle tissue.

Voluntary Muscle Tissue

Voluntary muscle tissue is made up of bands of elastic or contractile tissue bound together in bundles and enclosed by a connective tissue sheath called a **fascia**. This protects the muscle and helps to give it a contoured shape. The ends of the sheath extend to form tendons, by which voluntary muscles are attached to bone.

KEY NOTE – TENDON

A tendon is made up of white fibrous tissue, which attaches the muscle to the vascular membrane of a bone.

Voluntary muscles have many nuclei situated on their outer membrane. In microscopic structure, they are known to have a large number of striated fibres; this is because the contractile fibres that form them are connected in such a way that they appear to be striped. These contractile fibres or myofibrils in skeletal muscle run longitudinally and consist of two kinds of protein filaments:

▶ actin – the thinner filament
▶ myosin – the thicker filament.

The two types of filaments are arranged in alternating bands, hence they appear *striped* or *striated*. These protein filaments are significant in the mechanism of muscle contraction.

Voluntary muscle works intimately with the nervous system and will therefore only contract if a stimulus is applied to it via a motor nerve. Each muscle fibre receives its own nerve impulse so that fine and varied motions are possible. Voluntary muscles also have their own small stored supply of glycogen which is used as fuel for energy. Voluntary muscle tissue differs from other types of muscle tissue in that the muscles tire easily, and need regular exercise.

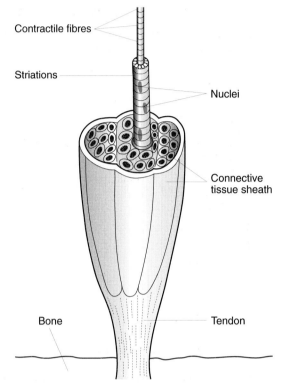

Figure 33
Structure of voluntary muscle tissue

Cardiac Muscle

Cardiac muscle is a specialised type of involuntary muscle tissue found only in the walls of the heart. Forming the bulk of the wall of each heart chamber, cardiac muscle contracts rhythmically and continuously to provide the pumping action necessary to maintain a relatively consistent flow of blood throughout the body.

Cardiac muscle resembles voluntary muscle tissue in that it is striated due to the actin and myosin filaments. However, it differs in that it:

▶ is branched in structure
▶ has intercalated discs in between each cardiac muscle cell, which form strong junctions to assist in the rapid transmission of impulses throughout an entire section of the heart, rather than in bundles.

The contraction of the heart is automatic; the stimulus to contract is stimulated from a specialised area of muscle in the heart called the **sinoatrial** (SA) **node**, which controls the heart rate.

As the heart has to alter its force of contraction due to regional requirements, its contraction is regulated not only by nerves, but also by hormones. For example, adrenaline in the blood can speed up contractions.

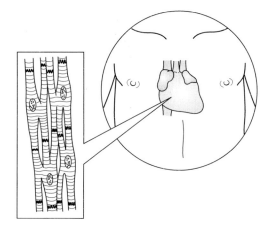

Figure 34
Cardiac muscle (found in the wall of the heart)

Smooth Muscle

Smooth muscle is also known as *involuntary* muscle, as it is not under the control of the conscious part of the brain. It is found in the walls of hollow organs such as the stomach, intestines, bladder, uterus and in blood vessels.

The main characteristics of smooth muscle are:

▶ the muscle cells are spindle-shaped and tapered at both ends
▶ each muscle cell contains one centrally located oval-shaped nucleus.

Smooth muscle has no striations due to the different arrangement of the protein filaments actin and myosin which are attached at their ends to the cell's plasma membrane.

The muscle fibres of smooth muscle are adapted for long, sustained contraction and therefore consume very little energy. One of the special features of smooth muscle is that it can stretch and shorten to a greater extent and still maintain its contractile function. Smooth muscle will contract or relax in response to nerve impulses, stretching or hormones but it is not under voluntary control.

Smooth muscle, like voluntary muscle has muscle tone and this is important in areas such as the intestines where the walls have to maintain a steady pressure on the contents.

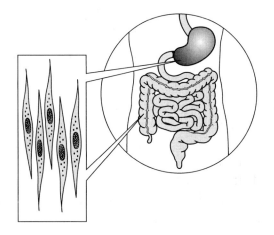

Figure 35
Smooth muscle (found in the wall of the stomach and intestines)

Muscle Contraction

Muscle tissue has several characteristics which help contribute to the functioning of a muscle:

▶ *Contractibility*: the capacity of the muscle to shorten and thicken
▶ *Extensibility*: the ability to stretch when the muscle fibres relax
▶ *Elasticity*: the ability to return to its original shape after contraction
▶ *Irritability*: the response to stimuli provided by nerve impulses.

Muscles vary in the *speed* at which they contract. The muscle in your eyes will be moving very fast as you are reading this page, whilst the muscles in your limbs assisting you in turning the pages will be contracting at a moderate speed.

The speed of a muscle contraction is therefore modified to meet the demands of the action concerned and the degree of nervous stimulus it has received.

Skeletal or voluntary muscles are moved as a result of nervous stimulus which they receive from the brain via a motor nerve. Each nerve fibre ends in a motor point, which is the end portion of the nerve, and is the part through which the stimulus to contract is given to the muscle fibre. A single motor nerve may transmit stimuli to one muscle fibre or as many as 150, depending on the effect of the action required. Cardiac and smooth muscle are innervated by the autonomic nervous system.

The Contraction of Voluntary Muscle Tissue

The functional characteristic of muscle is its ability to transform chemical energy into mechanical energy in order to exert force.

When a stimulus is applied to voluntary muscle fibres via a motor nerve a *mechanical* action is initiated:

▶ During contraction a sliding movement occurs within the contractile fibres of the muscle in which the actin protein filaments move inwards towards the myosin and the two filaments merge. This action causes the muscle fibres to shorten and thicken and then pull upon their attachments (bones and joints) to effect the movement required. The attachment of myosin cross-bridges to actin requires the mineral calcium. The nerve impulses leading to contraction cause an increase in calcium ions within the muscle cell.

▶ During relaxation, the muscle fibres elongate and return to their original shape

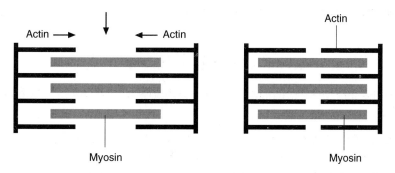

Figure 36
Actin and myosin

The force of muscle contraction depends upon the number of fibres in a muscle which contract simultaneously, and the more fibres involved, the stronger and more powerful the contraction will be.

KEY NOTE

The basic contractile process is the same in cardiac, smooth and voluntary muscles, with movement being achieved through the action of the protein filaments actin and myosin. However, since the requirements are different in terms of speed and force of contraction, the structure of cardiac and smooth muscles are slightly different to voluntary muscle tissue.

The Energy Needed for Muscle Contraction

A certain amount of energy is needed to effect the *mechanical* action of the muscle fibres. This is obtained principally from carbohydrate foods such as glucose in the arterial blood supply. Glucose, which is not required immediately by the body, is converted into glycogen and is stored in the liver and the muscles. Muscle glycogen therefore provides the *fuel* for muscle contraction.

▶ During muscle contraction glycogen is broken down by a process called oxidation (where glucose combines with oxygen and releases energy). Oxygen is stored in the form of haemoglobin in the red blood cells and as myoglobin in the muscle cells.

▶ During oxidation, a chemical compound called ATP (adenosine triphosphate) is formed. Molecules of ATP are contained within voluntary muscle tissue and their function is to temporarily store energy produced from food.

▶ When the muscle is stimulated to contract, ATP is converted to another chemical compound, ADP (adenosine diphosphate), which releases the energy needed to be used during the phase of muscle contraction.

▶ During the oxidation of glycogen, a substance called pyruvic acid is formed.

▶ If plenty of oxygen is available to the body, as in rest or undertaking moderate exercise, then the pyruvic acid is broken down into waste products, carbon dioxide and water which are excreted into the venous system. This is known as *aerobic* respiration.

▶ If insufficient oxygen is available to the body as may be in the case of vigorous exercise then the pyruvic acid is converted into lactic acid. This is known as *anaerobic* respiration.

> ## KEY NOTE
>
> The waste product lactic acid, which diffuses into the bloodstream after vigorous exercise, causes the muscles to ache. This condition is known as muscle fatigue which is therefore defined as: the loss of the ability of a muscle to contract efficiently due to insufficient oxygen, exhaustion of energy supply and the accumulation of lactic acid.

The Effects of Increased Circulation on Muscle Contraction

During exercise, muscles require more oxygen to cope with the increased demands made on the body. The body is then active in initiating certain circulatory and respiratory changes to the body to meet the increased oxygen requirements of the muscles.

Circulatory Changes that Occur in the Body during Muscle Contraction

▶ During exercise, there is an increased return of venous blood to the heart, owing to the more extensive movements of the diaphragm and the general contractions of the muscles compressing the veins.

▶ With the rate and output of each heart beat being increased, a greater volume of blood is circulated around the body, which will lead to an increase in the amount of oxygen in the blood.

▶ More blood is distributed to the muscle and less to the intestine and skin to meet the needs of the exercising muscles. During exercise, a muscle may receive as much as 15 times its normal flow of blood.

Respiratory Changes

▶ The presence of lactic acid in the blood stimulates the respiratory centre in the brain, increasing the rate and depth of breath, producing panting.

The rate and depth of breath remains above normal for a while after strenuous exercise has ceased; large amounts of oxygen are taken in to allow the cells of the muscles and the liver to dispose of the accumulated lactic acid by oxidising it and converting it to glucose or glycogen. Lactic acid is formed in the tissues in amounts far greater than can be immediately disposed of by available oxygen. The extra oxygen needed to remove the accumulated lactic acid is what is called the oxygen debt, which must be repaid after the exercise is over.

KEY NOTE

The conversion of lactic acid back into glucose is a relatively slow process and it may take several hours to repay the oxygen debt, depending on the extent of the exercise undertaken.

This situation can be minimised by massaging muscles before and after an exercise schedule, which will increase the blood supply to the muscles and prevent an excess of lactic acid forming in the muscles.

The Effects of Temperature on Muscle Contraction

Exercising muscles produces heat, which is carried away from the muscle by the bloodstream and is distributed to the rest of the body. Exercise is, therefore, an effective way to increase body temperature. When muscle tissue is warm, the process of contraction will occur faster due to the acceleration of the chemical reactions and the increase in circulation. However, it is possible for heat cramps to occur in muscles which are exercised at high temperatures, as increased sweating causes loss of sodium in the body, leading to a reduction in the concentration of sodium ions in the blood supplying the muscle.

Cramp occurs when muscles become over-contracted and hence go into spasm; this is usually caused by an irritated nerve or an imbalance of mineral salts such as sodium in the body. Cramp most commonly affects the calf muscles or the sole of the feet. Cramp can be very painful as it is a sudden involuntary contraction of the muscle.

▶ Treatment to relieve the pain of cramp includes stretching the affected muscle group and using soothing effluerage movements to help to relax the muscles.

Conversely, as muscle tissue is cooled, the chemical reactions and circulation slow, causing the contraction to be slower. This causes an involuntary increase in muscle tone known as shivering, that increases body temperature in response to cold.

Muscle Tone

Even in a relaxed muscle, a few muscle fibres remain contracted to give the muscle a certain degree of firmness. At any given time, a small number of motor units in a muscle are stimulated to contract and cause tension in the muscle rather than full contraction and movement, whilst the others remain relaxed. The group of motor units functioning in this way change periodically so that muscle tone is maintained without fatigue. This state of partial contraction of a muscle is known as muscle tone and is important for maintaining body posture.

▶ Good muscle tone may be recognised by the muscles appearing firm and rounded
▶ poor muscle tone may be recognised by the muscles appearing loose and flattened.

KEY NOTE

An increase in the size and diameter of muscle fibres, usually caused by exercise and weight lifting, leads to a condition called hypertrophy.

Muscles with less than the normal degree of tone are said to be *flaccid* and when the muscle tone is greater than normal the muscles become *spastic* and rigid.

KEY NOTE

Muscle tone will vary from client to client and will largely depend on the amount of exercise undertaken. Muscles with good tone have a better blood supply as their blood vessels will not be inhibited by fat.

Disorders of the Muscular System

Fibromyalgia

A chronic condition that produces musculo-skeletal pain. Predominant symptoms include widespread musculoskeletal pain, lethargy and fatigue. Other characteristic features include a non-refreshing sleep pattern in which the patient feels exhausted and more tired than later in the day, and interrupted sleep.

▶ Other recognised symptoms include early morning stiffness, pins and needles sensation, unexplained headaches, poor concentration, memory loss, low mood, urinary frequency, abdominal pain, irritable bowel syndrome.
▶ Anxiety and depression are also common.

Fibrositis

An inflammatory condition of the fibrous connective tissues, especially in the muscle fascia (also known as muscular rheumatism).

Muscular atrophy

This is the wasting away of muscles due to poor nutrition, lack of use or may be due to a dysfunction of the motor nerve impulses.

Muscle spasm

An increase in muscle tension due to excessive motor nerve activity resulting in a knot in the muscle.

Muscle cramp

This is an acute painful contraction of a single muscle or group of muscles. Cramp is often associated with a mineral deficiency, an irritated nerve or muscle fatigue.

Muscular Dystrophy

A progressively crippling disease in which the muscles gradually weaken and atrophy. The cause is unknown.

Shin Splints

A soreness in the front of the lower leg due to straining of the flexor muscles used in walking.

Spasticity

This is characterised by an increase in muscle tone and stiffness. In severe cases, movements may become uncoordinated and involve a nervous dysfunction. Spasticity involves muscles with excessive tone, and is a condition often associated with nervous dysfunction.

Strain

A strain is an injury that is caused by excessive stretching or working of a muscle or tendon that results in a partial or complete tear.

▶ Symptoms include pain, swelling, tenderness and stiffness in the affected area.
▶ Muscle strains are more common in the lower back and the neck.

Sprain

A complete or incomplete tear in the ligaments around a joint. It usually follows a sudden, sharp twist to the joint that stretches the ligaments and ruptures some or all of its fibres.

▶ Sprains commonly occur in the ankle, wrist and the back where there is localised pain, swelling and loss of mobility.

Torticollis

A condition in which the neck muscles (sternomastoids) contract involuntarily. It is commonly called 'wryneck'.

Whiplash

A condition produced by damage to the muscles, ligaments and intervertebral discs or nerve tissues of the cervical region by sudden hyperextension and/or flexion of the neck.

▶ The most common cause is a road traffic accident when acceleration/deceleration causes sudden stretch of the tissue around the cervical spine. It may also occur as a result of hard impact sports.
▶ It can present with pain, limitation of neck movements with muscle tenderness, which can start hours to days after the accident, and may take months to recover. This is usually affected by complicated physical and psychological issues.

SELF-ASSESSMENT QUESTIONS – THE MUSCULAR SYSTEM

1. State the three main types of muscle tissue, stating where they are found in the body.

2. Briefly state the structure of the three main types of muscle tissue in the body.

3. State the structure and function of a tendon.

4. Describe how a muscle contracts.

....

....

....

....

5. State the body's principal fuel for muscle contraction.

....

....

6. What is meant by the term muscle fatigue? How can this be alleviated?

....

....

....

7. What is meant by the term muscle tone?

....

8. How would you recognise the following:

a) good muscle tone?

....

b) poor muscle tone?

....

9. Briefly explain the effects of increased blood circulation on muscle contraction.

....

....

....

10. Briefly explain the effects of temperature on muscle contraction.

....

....

....

Muscles of the Head and Neck
Frontalis

▶ **Position and attachments** – this muscle extends over the front of the skull and the width of the forehead. It attaches to the skin of the eyebrows and the frontal bone at the hairline.

▶ **Action** – wrinkles the forehead and raises the eyebrows.

KEY NOTE

This muscle is used when expressing surprise.

Occipitalis

▶ **Position and attachments** – this muscle is found at the back of the head. It is attached to the occipital bone and the skin of the scalp.

▶ **Action** – moves the scalp backwards.

KEY NOTE

This muscle is united to the frontalis muscle by a broad tendon called the **epicranial aponeurosis**, which covers the skull like a cap.

Temporalis

▶ **Position and attachments** – this is a fan-shaped muscle situated on the side of the skull above and in front of the ear. It attaches to the temporal bone and to the upper part of the mandible.

▶ **Action** – raises the lower jaw when chewing.

> ### *KEY NOTE*
>
> This muscle becomes over-tight and painful in the condition known as Temporomandibular joint dysfunction syndrome.

Orbicularis Oculi

▶ **Position and attachments** – this is a circular muscle that surrounds the eye. It is attached to the bones at the outer edge and the skin of the upper and lower eyelids at the inner edge.
▶ **Action** – closes the eye.

> ### *KEY NOTE*
>
> This muscle is used when blinking or winking. It also compresses the lacrimal gland, aiding the flow of tears.

Orbicularis Oris

▶ **Position and attachments** – a circular muscle that surrounds the mouth. Its fibres attach to the maxilla, mandible, the lips and the buccinator muscle.
▶ **Action –** closes the mouth.

> ### *KEY NOTE*
>
> This muscle is used when shaping the lips for speech and when kissing. It also contracts the lips when tense.

Corrugator

▶ **Position and attachments** – this muscle is located in between the eyebrows. It is attached to the frontalis muscle and the inner edge of the eyebrow.
▶ **Action** – brings the eyebrows together.

> **KEY NOTE**

This muscle is used when frowning.

Procerus

▶ **Position and attachments** – this muscle in located in between the eyebrows. It is attached to the nasal bones and the frontalis muscle.
▶ **Action** – draws the eyebrows inwards.

> **KEY NOTE**

The contraction of the muscle creates a puzzled expression.

Nasalis

▶ **Position and attachments** – this muscle is located at the sides of the nose. It is attached to the maxillae bones and the nostrils.
▶ **Action** – dilates and compresses the nostrils.

> **KEY NOTE**

This muscle is used when blowing the nose.

Zygomatic Major and Minor / Zygomaticus

▶ **Position and attachments** – lies in the cheek area, extending from the zygomatic bone to the angle of the mouth.
▶ **Action** – draws the angle of the mouth upward and laterally.

> **KEY NOTE**

This muscle is used when laughing and smiling.

Levator Labii Superioris

▶ **Position and attachments** – this muscle is located towards the inner cheek beside the nose and extends from the upper jaw to the skin of the corners of the mouth and the upper lip.

▶ **Action** – raises the upper lip and the corner of the mouth.

KEY NOTE

This muscle is used to create a snarling expression.

Risorius

▶ **Position and attachments** – a triangular-shaped muscle that lies horizontally on the cheek, joining at the corners of the mouth. Risorius lies above the Buccinator muscles (see below) and is attached to the zygomatic bone at one end and the skin of the corner of the mouth at the other.

▶ **Action** – pulls the corner of the mouth sideways and upwards.

KEY NOTE

This muscle creates a grinning expression.

Buccinator

▶ **Position and attachments** – this muscle is the main muscle of the cheek. It is attached to both the upper and lower jaw: its fibres are directed forward from the bones of the jaws to the angle of the mouth.

▶ **Action** – this muscle helps hold food in contact with the teeth when chewing and compresses the cheek.

KEY NOTE

This muscle is used when blowing a balloon or blowing a trumpet. It is also a common site for holding tension in the face.

Mentalis

▶ **Position and attachments** – this muscle radiates from the lower lip over the centre of the chin. It is attached to the lower jaw and the skin of the lower lip.
▶ **Action** – elevates the lower lip and wrinkles the skin of the chin.

KEY NOTE

This muscle is used when expressing displeasure and when pouting.

Masseter

▶ **Position and attachments** – this is a thick, flattened, superficial muscle; its fibres extend downwards from the zygomatic arch to the mandible.
▶ **Action –** the main muscle of mastication. It raises the jaw and exerts pressure on the teeth when chewing.

KEY NOTE

This muscle holds a lot of tension and can be felt just in front of the ear when the teeth are clenched.

Triangularis

▶ **Position and attachments** – a triangular-shaped muscle, located under the corners of the mouth. Its fibres stretch from the lower jaw to the skin and muscles of the corner of the mouth.
▶ **Action** – draws the corners of the mouth downwards.

KEY NOTE

The contraction of this muscle creates an expression of sadness.

Sternomastoid

▶ **Position and attachments** – long muscle that lies obliquely across each side of the neck. Its fibres extend upwards from the sternum and clavicle at one end to the mastoid process of the temporal bone (at the back of the ear).

▶ **Action** – when working together they flex the neck (pull the chin down towards the chest) and when working individually, they rotate the head to the opposite side.

KEY NOTE

Spasm of the sternomastoid muscle results in a condition known as Torticollis or wryneck. Sternomastoid is the only muscle that moves the head but is not attached to any vertebrae.

Platysma

▶ **Position and attachments –** a superficial neck muscle that extends from the chest (fascia covering the upper part of pectoralis major and deltoid) up either side of the neck to the chin.

▶ **Action** – depresses the lower jaw and lower lip.

KEY NOTE

This muscle is used in yawning and when creating a pouting expression.

TASK 1

Label the muscles of the head and neck on Figure 37:

frontalis	temporalis	orbicularis oris	nasalis
corrugator	zygomatic major	orbicularis oculi	procerus
zygomatic minor	levator labii superioris	risorius	buccinator
mentalis	masseter	platysma	triangularis
sternamastoid			

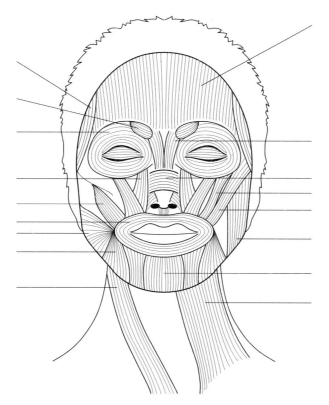

Figure 37
Muscles of the head and neck

SELF-ASSESSMENT QUESTIONS –
MUSCLES OF THE HEAD AND NECK

1. Identify the following muscles from the description given

a) the muscle that closes the mouth

b) the muscle that turns the head to the opposite side

c) the muscle that draws the corners of the mouth upwards

d) the muscle that raises the upper lip

e) the muscle that wrinkles the forehead and raises the eyebrows

f) the muscle that exerts pressure on the teeth when chewing

2. Give the action of the following muscles

a) Mentalis

b) Buccinator

c) Risorius

d) Orbicularis oculi

e) Platsyma

Muscles of the Shoulder
Trapezius

▶ **Position** – a large triangular-shaped muscle in the upper back that extends horizontally from the base of the skull (occipital bone) and the cervical and thoracic vertebrae to the scapula.
Its fibres are arranged in three groups: upper, middle and lower.

▶ **Action** – The upper fibres raise the shoulder girdle; the middle fibres pull the scapula towards the vertebral column and the lower fibres draw the scapula and shoulder downward. When the trapezius is fixed in position by other muscles, it can pull the head backwards or to one side.

KEY NOTE

This muscle tends to hold a lot of upper body tension, causing discomfort and restrictions in the neck and shoulder.

Levator Scapula

▶ **Position** – a strap-like muscle that runs almost vertically through the neck, connecting the cervical vertebrae to the scapula.

▶ **Action** – elevates and adducts the scapula.

KEY NOTE

This muscle tends to become very tight, affecting mobility of the neck and shoulder.

Rhomboids

▶ **Position** – the fibres of these muscles lie between the scapulae. They attach to the upper thoracic vertebrae at one end and the medial border of the scapula at the other end.

▶ **Action** – adduct the scapula.

> ### *KEY NOTE*
>
> These muscles are often very tight, resulting in aching and soreness in between the scapulae.

Supraspinatus

▶ **Position** – this muscle is located in the depression above the spine of the scapula. It is attached to the spine of the scapula at one end and the humerus at the other.
▶ **Action** – abducts humerus, assisting the deltoid.

> ### *KEY NOTE*
>
> This muscle often becomes fatigued when working for prolonged periods at a desk or computer, or when driving.

Infraspinatus

▶ **Position** – this muscle attaches to the middle two-thirds of the scapula below the spine of the scapula at one end, and the top of the humerus at the other.
▶ **Action** – rotates humerus laterally (outwards).

> ### *KEY NOTE*
>
> Tension in this muscle can affect the range of mobility in the shoulder.

Teres Major

▶ **Position** – this muscle attaches to the bottom lateral edge of the scapula at one end and the back of the humerus (just below the shoulder joint) at the other.
▶ **Action** – adducts and medially (inwardly) rotates humerus.

> **KEY NOTE**
>
> Tension in this muscle restricts the mobility of the shoulder and upper arm.

Teres Minor

▸ **Position** – this muscle attaches to the lateral edge of the scapula, above teres major at one end, and into the top of the posterior of the humerus at the other.
▸ **Action** – rotates humerus laterally (outwards).

> **KEY NOTE**
>
> Tension in this muscle restricts the mobility of the shoulder and upper arm.

Subscapularis

▸ **Position** – this muscle attaches to the inside surface of the scapula to the anterior of the top of the humerus.
▸ **Action** – rotates the humerus medially, draws the humerus forward and down when the arm is raised.

> **KEY NOTE**
>
> This muscle is often implicated in the case of a frozen shoulder.

Deltoid

▸ **Position** – a thick triangular muscle that caps the top of the humerus and shoulder. It attaches to the clavicle and the spine of the scapula at one end, and to the side of the humerus at the other.
▸ **Action** – abducts arm, draws the arm backwards and forwards.

> **KEY NOTE**
>
> This muscle tends to hold upper body tension and will often go into spasm, along with the trapezius muscle.

Coraco-brachialis

▶ **Position** – extends from the scapula to the middle of the humerus along its medial surface.

▶ **Action** – flexes and adducts the humerus.

TASK 2

Label the following muscles on Figure 38:

trapezius   levator scapula   teres minor   rhomboids   supraspinatus

infraspinatus   teres major   subscapularis   deltoid   caracobrachialis

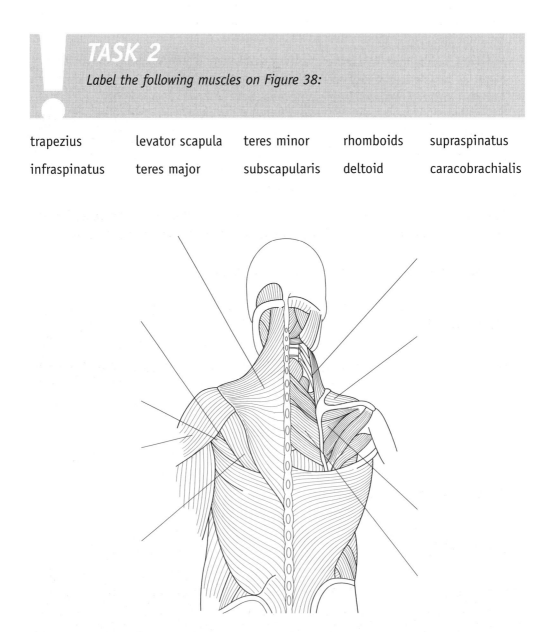

Figure 38
Muscles of the shoulder

Muscles of the Upper Limbs

Biceps

▶ **Position** – anterior of the upper arm (humerus). It attaches to the scapula at one end and the radius and flexor muscles of the forearm at the other.

▶ **Action** – flexes the forearm at the elbow; supinates the forearm.

KEY NOTE

The actions of the biceps muscle are likened to the action of removing a corkscrew from a wine bottle.

Triceps

▶ **Position** – posterior of the upper arm (humerus). It attaches to the posterior of the humerus and outer edge of the scapula at one end, and to the ulna below the elbow at the other.

▶ **Action** – extension (straightening) of the forearm.

KEY NOTE

The triceps is also referred to as the boxer's muscle as it is used when delivering a 'knock-out' punch.

Brachialis

▶ **Position** – lies beneath biceps. It attaches to the distal half of the anterior surface of the humerus at one end and the ulna at the other.

▶ **Action** – flexes the forearm at the elbow.

KEY NOTE

The brachialis is the most effective flexor muscle of the forearm due to its anatomical position, lying underneath the biceps.

Brachioradialis

▶ **Position** – this muscle connects the humerus to the radius; it attaches to the distal end of the humerus at one end and the radius at the other end.
▶ **Action** – flexes forearm at the elbow.

KEY NOTE

The brachioradialis can be felt as the bulge on the radial side of the forearm.

Pronator Teres

▶ **Position** – crosses the anterior aspect of the elbow. It attaches to the distal end of the humerus and the upper aspect of the ulna at one end, and the lateral surface of the radius at the other.
▶ **Action** – pronates and flexes forearm.

Supinator

▶ **Position** – attaches to the lateral aspect of the lower humerus and the radius.
▶ **Action** – supinates the forearm.

Flexors of the Forearm

▶ **Position** – lie on the medial aspect of the forearm attached to the lower part of the humerus, radius and ulna at one end, and the metacarpals and phalanges of the fingers at the other end.
▶ **Action** – flex the wrist, fingers and thumbs.

KEY NOTE

These muscles can become easily inflamed due to excess pressure and overwork (as in typing, working on a keyboard).

Extensors

▶ **Position** – lie on the lateral aspect of the forearm attached to the lower humerus, radius and ulna at one end, and the metacarpals and phalanges at the other.

▶ **Action** – extend the wrist, fingers and thumb.

> ### KEY NOTE
>
> These muscles can become easily inflamed due to excess pressure and overwork (as in typing, working on a keyboard).

TASK 3

Label the following muscles on Figure 39:

biceps   triceps   brachialis   brachioradialis   pronator teres   flexors of forearm
extensors of forearm

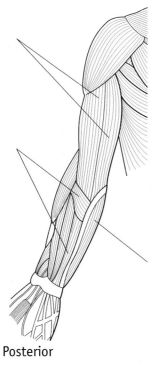
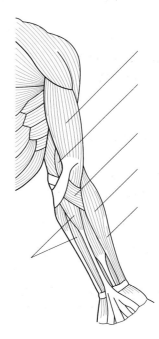

Posterior   Anterior

Figure 39
Muscles of the upper limb

Muscles of the Lower Limb
Quadriceps Extensor

The quadriceps is made up of four muscles: Rectus Femoris, Vastus Lateralis, Vastus Intermedius, Vastus Medialis.

- **Position** – anterior aspect of the thigh attached to the pelvic girdle (rectus femoris) and femur (vastus group) at one end to the patella, and tibia at the other end.
- **Action** – as a group they extend the knee and flex the hip.

KEY NOTE

The quadriceps muscles are used when walking, kicking and raising the body from a sitting or squatting position.

Sartorius

- **Position and attachments** – attached to the ilium of the pelvis, this muscle crosses the anterior of the thigh to the medial aspect of the tibia.
- **Action** – flexes the hip and knee; rotates the thigh laterally (turns it outwards).

KEY NOTE

Due to its unusual position, the sartorius can flex both the hip and the knee. Sartorius can lead to knee problems, because turning the leg outwards puts pressure on the knee.

Adductors

This is a group of four muscles: Adductor Brevis, Adductor Longus, Adductor Magnus, Pectineus

- **Position and attachments** – these muscles are situated on the medial aspect of the thigh. They are attached to the lower part of the pelvic girdle at one end (pubic bones and the ischium), and the inside of the femur at the other end.
- **Action** – as a group they adduct and laterally rotate the thigh. They also flex the hip.

KEY NOTE

The adductors are important muscles in the maintenance of body posture. Groin strains are a common problem associated with these muscles.

Gracilis

▶ **Position and attachments** – a long strap-like muscle attached to the lower edge of the pubic bone at one end and the upper part of the medial aspect of the tibia at the other end.
▶ **Action** – adducts thigh, flexes knee and hip, medially (inwardly) rotates the thigh and tibia.

Hamstrings

The hamstrings consist of three muscles: two situated on the inside of the thigh (semitendinosus and semimembranosus) and one on the outside of the thigh (biceps femoris).

▶ **Position and attachments** – posterior aspect of the thigh attaches to the lower part of the pelvis (ischium), and lower part of the posterior of the femur to either side of the posterior of the tibia.
▶ **Action** – flex the knee and extend the hip.

KEY NOTE

The hamstring muscles contract powerfully when raising the body from a stooped position, and when climbing stairs.

Tensor Fascia Latae

▶ **Position and attachments** – runs laterally down the side of the thigh. It is attached to the outer edge of the ilium of the pelvis and runs via the long fascia lata tendon to the lateral aspect of the top of the tibia.
▶ **Action** – flexes, abducts and medially rotates thigh.

KEY NOTE

Tensor fascia latae is attached to a broad sheet of connective tissue (fascia lata tendon), which helps to strengthen the knee joint when walking and running.

Gastrocnemius

▶ **Position and attachments** – large superficial calf muscle with two bellies (central portion of the muscle) on the posterior of the lower leg. Attached to the lower aspect of the posterior of the femur across the back of the knee, and runs via the achilles tendon ankle to the calcaneum at the back of the heel.
▶ **Action** – plantar flexes the foot; assists in knee flexion.

KEY NOTE

Gastrocnemius provides the push during fast walking and running.

Soleus

▶ **Position** – situated deep to gastrocnemius in the calf, soleus is attached to the tibia and fibula just below the back of the knee at one end, and runs via the achilles tendon to the calcaneum at the other end.
▶ **Action** – plantar flexes the foot.

KEY NOTE

The soleus is flatter and thicker than the gastrocnemius. It is important as a postural muscle.

Tibialis Anterior

▶ **Position** – anterior aspect of the lower leg, attached to the outer side of the tibia at one end and the medial cuneiform and the base of the first metatarsal at the other end.
▶ **Action** – dorsiflexes and inverts the foot.

KEY NOTE

If this muscle becomes weak, it can lead to the lower leg rolling inwards due to the collapse of the medial longitudinal arch of the foot.

Tibialis Posterior

▶ **Position** – posterior aspect of the lower leg, very deeply situated in the calf. It is attached to the back of the tibia and fibula at one end and the navicular, third cuneiform and second, third and fourth metatarsals at the other.
▶ **Action** – assists in plantar flexion and inverts the foot.

KEY NOTE

Weakness in this muscle can cause the feet to turn out from the ankles rather than the knees; this causes the muscle to stretch and the medial longitudinal arch of the foot to drop.

Peroneus Longus / Brevis

▶ **Position and attachments –** situated on the lateral aspect of the lower leg, they attach to the fibula to the underneath of the first (longus) and fifth metatarsal (brevis).
▶ **Action** – plantar flexes and everts the foot.

KEY NOTE

Going over onto the outside of the ankle as in a trip or a fall can sprain these muscles. If the injury is not treated properly, it can affect future stability of the ankle joint.

Flexors of the Toes

▶ **Position and attachments** – muscles lying deep in the posterior aspect of the lower leg, attached to the tibia and fibula at one end and the phalanges of the toes at the other end.

▶ **Action** – flex the toes and help plantar flex the ankle.

KEY NOTE

These muscles can become weak due to excess pressure and overuse in walking and running.

Extensors of the Toes

▶ **Position and attachments** – muscles situated on the anterior and lateral aspects of the lower leg, attached to the tibia and fibula and the phalanges of the toes.
▶ **Action** – extend the toes and help dorsiflex the ankle.

KEY NOTE

These muscles can become weak due to excess pressure and overuse in walking and running.

Piriformis

▶ **Position and attachments** – a deeply-seated pelvic girdle muscle that attaches to the anterior of the sacrum at one end and the top of the femur at the other.
▶ **Action** – lateral rotation and abduction of the hip.

KEY NOTE

This is the largest of the lateral rotators of the hip. If it becomes over-tense, it restricts mobility in the hip.

TASK 4

Label the following muscles on Figure 40:

quadriceps extensor (vastus lateralis, vastus intermedius, vastus medialis, rectus femoris) adductors (brevis, longus, magnus, pectineus) gracilis sartorius hamstrings (biceps femoris, semimembranosus) tenser fascia lata gastrocnemius soleus tibialis anterior tibialis posterior peroneus longus

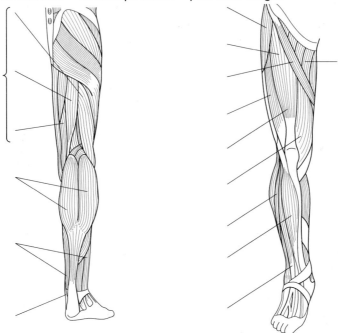

Figure 40
Muscle of the lower limb

Muscles of the Pelvic Floor

The **levator ani** and the **coccygeus** are the muscles that form the pelvic floor. These muscles support and elevate the organs of the pelvic cavity such as the uterus and the bladder. They provide a counterbalance to increased intra-abdominal pressure, which would expel the contents of the bladder, rectum and the uterus.

During childbirth these muscles can become weakened and need to be strengthened by pelvic floor exercises as soon as possible after the birth.

Muscles of the Anterior Aspect of the Trunk
Pectoralis Major

▶ **Position** – a thick fan-shaped muscle covering the anterior surface of the upper chest. It attaches to the clavicle and the sternum at one end and to the humerus at the other end.

▶ **Action** – adducts arm, medially (inwardly) rotates arm.

KEY NOTE

Tightness in this muscle may cause constriction of the chest or postural distortions (such as rounded shoulders).

Pectoralis Minor

▶ **Position** – a thin muscle that lies beneath the pectoralis major. Its fibres attach laterally and upwards from the ribs at one end to the scapula at the other end.

▶ **Action** – draws the shoulder downwards and forwards.

KEY NOTE

This muscle assists in forced respiration and is therefore an accessory respiratory muscle.

Serratus Anterior

▶ **Position** – a broad curved muscle located on the side of the chest / rib cage below the axilla. It attaches to the outer surface of the upper eighth or ninth rib at one end to the inner surface of the scapula, along the medial edge nearest the spine.

▶ **Action** – pulls the scapula downwards and forwards.

KEY NOTE

This muscle is used to thrust the shoulder forward as in pushing. It is often referred to as the boxer's muscle as its action enables a boxer to deliver a punch.

External Obliques

▶ **Position** – a broad, thin sheet of muscle whose fibres slant downwards from the lower ribs to the pelvic girdle and the linea alba (tendon running from the bottom of the sternum to the pubic symphysis).

▶ **Action** – flexes, rotates and side-bends the trunk; compresses the contents of the abdomen.

KEY NOTE

The fibres of the external obliques run in the direction in which you put your hands in your pocket.

Internal Obliques

▶ **Position** – a broad, thin sheet of muscle located beneath the external obliques. Its fibres run up and forward from the pelvic girdle to the lower ribs.

▶ **Action** – flexes, rotates and side-bends the trunk; compresses the contents of the abdomen.

KEY NOTE

The fibres of the internal obliques run at right angles to the external obliques.

External Intercostals

▶ **Position** – these are the superficial muscles that occupy and attach to the space between the ribs (called external because they are positioned on the outside).

▶ **Action** – help to elevate the rib cage during inhalation.

KEY NOTE

The external intercostals help to increase the depth of the thoracic cavity.

Internal Intercostals

▶ **Position** – these muscles lie deep to the external intercostals (called internal because they are positioned on the inside). They occupy and attach to the spaces between the ribs.
▶ **Action** – depress the rib cage, which helps to move air out of the lungs when exhaling.

KEY NOTE

The internal intercostals are used during forced expiration (for example, when coughing).

Rectus Abdominis

▶ **Position** – a long strap-like muscle that attaches to the pubic bones at one end and the ribs and the sternum at the other.
▶ **Action** – flexes the vertebral column, flexes the trunk (as in a sit-up), compresses the abdominal cavity.

KEY NOTE

This muscle is crossed transversely by three or more fibrous bands that give it a segmented appearance (the so-called 'six pack'). It is an important postural muscle.

Diaphragm

▶ **Position** – a large dome-shaped muscle that separates the thorax from the abdomen. It attaches to the lower part of the sternum, lower six ribs and upper three lumbar vertebrae, and its fibres converge to meet on a central tendon in the abdominal cavity.
▶ **Action** – on contraction the diaphragm flattens to expand the volume of the thoracic cavity to assist inspiration. Upon relaxation and expiration it returns to its dome shape.

> ## KEY NOTE

The diaphragm is the chief muscle of respiration.

Transverse Abdominus / Transversalis

▶ **Position and attachments** – attaches to the inner surfaces of the ribs (last six) and iliac crest at one end, and extends down to the pubis via the linea alba (a long tendon that extends from the bottom of the sternum to the pubic symphysis).

▶ **Action** – compression of the abdominal contents; supports the organs of the abdominal cavity.

> ## KEY NOTE

The transverse abdominus is the deepest of the abdominal muscles: it wraps itself like a band around the internal organs in the abdomen.

TASK 5

Label the following muscles on Figure 41:

pectoralis major   pectoralis minor   external obliques   internal obliques

rectus abdominals   serratus anterior

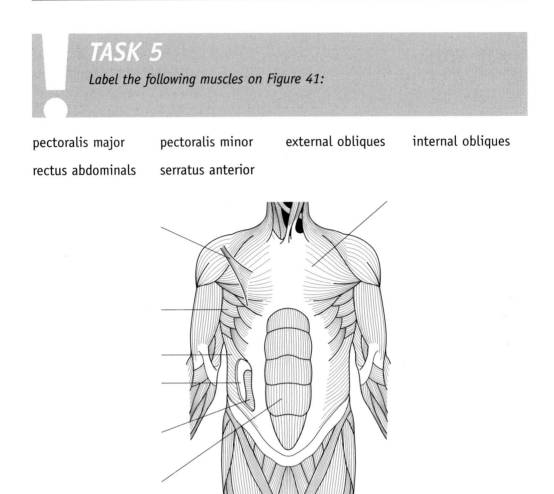

Figure 41
Muscles of the anterior aspect of the trunk

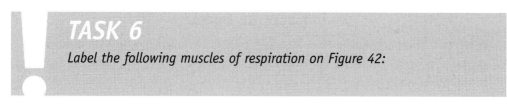

TASK 6

Label the following muscles of respiration on Figure 42:

diaphragm   external intercostals   internal intercostals

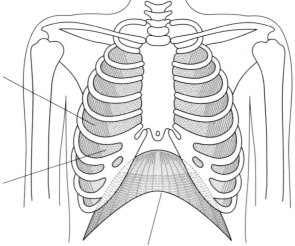

Figure 42
Muscles of respiration

Muscles of the Posterior Aspect of the Trunk

Erector Spinae / Sacrospinalis

▶ **Position** – this muscle is made up of separate bands of muscle that lie in the groove between the vertebral column and the ribs. They attach to the sacrum, iliac crest at one end to the ribs, transverse and spinous processes of the vertebrae and the occipital bone at the other.

▶ **Action** – extension, lateral flexion and rotation of the vertebral column.

KEY NOTE

This is a very important postural muscle as it helps to extend the spine.

Latissimus Dorsi

▶ **Position** – a broad muscle that attaches to the posterior of the iliac crest and sacrum, lower six thoracic and five lumbar vertebrae at one end and the humerus at the other end.

▶ **Action** – extends, adducts and rotates the humerus medially.

KEY NOTE

This muscle is often referred to as the 'swimmer's muscle' as it allows us to extend the arm to propel the body in water. It is one of the major muscles of the body implicated in lower back pain, due to its pelvic attachments.

Quadratus Lumborum

▶ **Position and attachments** – attaches to the top of the posterior of the iliac crest at one end to the twelfth rib and transverse processes of the first four lumbar vertebrae at the other end.

▶ **Action** – lateral flexion (side bending) of the lumbar vertebrae.

KEY NOTE

Excessive bending to the side can injure and strain the quadratus lumborum muscle.

Psoas

▶ **Position** – a long, thick deep pelvic muscle. It attaches to the anterior transverse processes of T12–L5 (twelfth thoracic to fifth lumbar vertebrae) to the inside of the top of the femur at the other end.

▶ **Action** – flexes the thigh.

Iliacus

▶ **Position** – a large fan-shaped muscle deeply situated in the pelvic girdle. It attaches to the iliac crest at one end, and to the inside of the top of the femur at the other end.

▶ **Action** – flexes and laterally rotates the femur.

The iliacus and psoas muscles are often considered as one, and are referred to as iliopsoas. They are the primary flexors of the thigh and they serve to advance the leg in walking.

Gluteus Maximus

▶ **Position and attachments** – a large muscle covering the buttock. It attaches to the back of the ilium along the sacroiliac joint at one end, and into the top of the femur at the other.

▶ **Action** – extends the hip and laterally rotates thigh.

Gluteus maximus is used when running and jumping. It is often implicated in postural defects such as lordosis, excess curvature in the lumbar spine.

Gluteus Medius

▶ **Position** – this muscle is partly covered by gluteus maximus. It attaches to the outer surface of the ilium at one end and the outer surface of the femur at the other end.

▶ **Action** – abducts thigh, medially rotates thigh.

The gluteus medius is used when walking and running. It is also used in balance and when standing on one leg.

Gluteus Mimimus

▶ **Position and attachments** – this muscle lies beneath the gluteus medius. Its attachments are the same as for gluteus medius: outer surface of the ilium at one end to the outer surface of the femur at the other end.

▶ **Action** – abducts thigh, medially rotates thigh.

KEY NOTE

The gluteus minimus is also used when walking and running, along with gluteus medius. It is also used in balance and when standing on one leg.

TASK 7

Label the following muscles on Figure 43:

erector spinae   latissimus dorsi   quadratus lumborum   psoas   iliacus

gluteus maximus   gluteus medius   gluteus mimimus

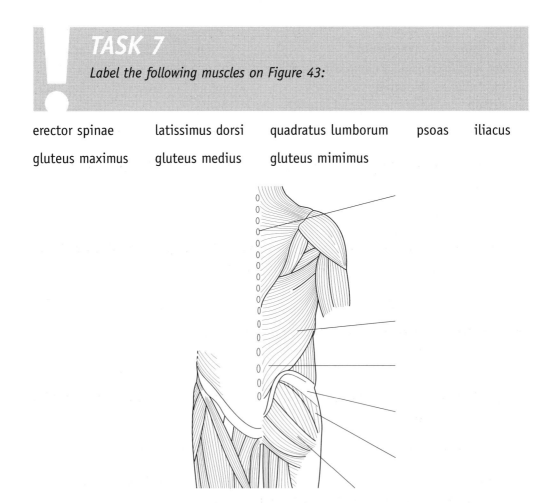

Figure 43
Muscles of the posterior aspect of the trunk

? SELF-ASSESSMENT QUESTIONS – MUSCLES OF THE BODY

1. Identify the following muscles from the description given

a) a thick fan-shaped muscle covering the anterior surface of the upper chest

b) a thick triangular muscle that caps the top of the humerus and shoulder

c) a broad curved muscle located on the side of the chest / rib cage below the axilla.

d) the muscle that runs down the lateral aspect of the thigh and is attached to a broad sheet of tendon that strengthens the knee

e) a superficial muscle of the calf that has two bellies

f) a broad curved shaped muscle located on the side of the chest, across the sides of the rib cage and below the axilla

g) the muscle that is made up of three separate bands of muscle that lie in the groove between the vertebral column and the ribs

h) the muscle in the abdominal wall that flexes the vertebral column and the trunk

i)   the muscles that abduct and medially rotate the thigh

....

j)   the muscle that lies under the biceps and is the most effective flexor muscle of
     the forearm

....

k)   the muscle of the posterior of the upper arm that extends the forearm

....

l)   the muscles that flex, rotate and side bend the trunk

....

2. Name the muscles involved in respiration.

....

....

3. Give the action of the following muscles

a) trapezius

....

b) gluteus maximus

....

c) latissimus dorsi

....

d) biceps

....

e) sartorius

....

f) tibialis anterior

....

g) soleus

....

h) pectoralis minor

....

i) hamstrings

....

CHAPTER 8

The Circulatory System

The circulatory system is the body's transport system and comprises blood, blood vessels and the heart. Blood provides the fluid environment for our body's cells, and is transported in specialised tubes called blood vessels. The heart acts like a pump which keeps the blood circulating around the body in a constant circuit.

A competent therapist needs to be able to understand:

▶ the principles of circulation, to understand the physiological effects of treatments and to be able to carry out treatments safely and effectively

By the end of this chapter, you will be able to relate the following knowledge to your practical work carried out in the salon:

▶ The structural and functional significance of the different type of blood cells
▶ the structural and functional differences between the different blood vessels
▶ the major blood vessels of the heart
▶ the pulmonary and systemic blood circulation
▶ blood pressure and the pulse rate
▶ circulatory disorders such as high or low blood pressure and varicose veins.

Blood

Blood is the fluid tissue or medium in which all materials are transported to and from individual cells in the body. Blood is therefore the chief transport system of the body.

The Percentage Composition of Blood

1 55 per cent of blood is fluid or plasma, which is a clear, pale yellow slightly alkaline fluid which consists of the following substances:

> 91 per cent of plasma is water
> 9 per cent remaining consists of dissolved blood proteins, waste, digested food materials, mineral salts and hormones.

2 45 per cent of blood is made up of the blood cells erythrocytes, leucocytes and thrombocytes.

Blood Cells

There are three types of blood cells:

1 Erythrocytes – red blood cells
2 Leucocytes – white blood cells
3 Thrombocytes – platelets

Erythrocytes

Erythrocytes are disc-shaped in structure and make up more than 90 per cent of the formed elements in blood. They are formed in red bone marrow and contain the iron-protein compound haemoglobin. Their function is to transport oxygen to the cells and carry carbon dioxide away from the cells.

Figure 44
An erythrocyte (red blood cell)

Leucocytes

Leucocytes are the largest of all the blood cells and appear white due to their lack of haemoglobin. There are two main categories of white blood cells:

1 **Granulocytes** – these account for about 75 per cent of white blood cells and can be further divided into: **Neutrophils**, **Eosonophils** and **Basophils**.

2 Agranulocytes – these can be divided into:

▶ **Lymphocytes** – these account for about 20 per cent of all white blood cells
▶ **Monocytes** – these account for about 5 per cent of white blood cells.

The main function of the white blood cells is to protect the body against infection and disease in a process known as **phagocytosis** (which means to engulf and ingest microbes, dead cells and tissue).

Figure 45
A monocyte (type of leucocyte)

Thrombocytes

Thrombocytes are also known as platelets. These are small fragments of cells and are the smallest cellular elements of the blood. They are very significant in the blood clotting process, as the platelets initiate the chemical reaction that leads to the formation of a blood clot.

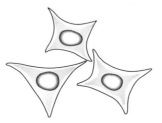

Figure 46
Thrombocyte (platelet)

Functions of Blood

There are four main functions of blood:

1 transport
2 defence
3 regulation
4 clotting

Transport

Blood is the primary transport medium for a variety of substances that travel throughout the body.

- Oxygen is carried from the lungs to the cells of the body in red blood cells.
- Carbon dioxide is carried from the body's cells to the lungs.
- Nutrients such as glucose, amino acids, vitamins and minerals are carried from the small intestine to the cells of the body.
- Cellular wastes such as water, carbon dioxide, lactic acid and urea are carried in the blood to be excreted.
- Hormones, which are internal secretions that help to control important body processes, are transported by the blood to target organs.

KEY NOTE

Red blood cells are called erythrocytes and they contain the red protein pigment haemoglobin that combines with oxygen to form oxyhaemoglobin. The pigment haemoglobin assists the function of the erythrocyte in transporting oxygen from the lungs to the body's cells and carrying carbon dioxide away.

Defence

- White blood cells are collectively called leucocytes and they play a major role in combating disease and fighting infection.

KEY NOTE

White blood cells are known as phagocytes as they have the ability to engulf and ingest micro-organisms which invade the body and cause disease. Specialised white blood cells called lymphocytes produce antibodies to protect the body against infection.

Regulation

- Blood helps to regulate heat in the body by absorbing large quantities of heat produced by the liver and the muscles; this is then transported around the body to help to maintain a constant internal temperature.
- Blood also helps to regulate the body's pH balance.

Clotting

▶ If the skin becomes damaged, specialised blood cells called platelets clot to prevent the body from losing too much blood and to prevent the entry of bacteria.

Blood Vessels

Blood flows round the body by the pumping action of the heart and is carried in vessels known as arteries, veins and capillaries.

Key Factors about Blood Vessels

Arteries

▶ Arteries carry blood away from the heart.
▶ Blood is carried under high pressure.
▶ Arteries have thick muscular and elastic walls to withstand pressure.
▶ Arteries have no valves, except at the base of the pulmonary artery, where they leave the heart.
▶ Arteries carry oxygenated blood, except the pulmonary artery to the lungs.
▶ Arteries are generally deep-seated, except where they cross over a pulse spot.
▶ Arteries give rise to small blood vessels called arterioles, which deliver blood to the capillaries.

Veins

▶ Veins carry blood towards the heart.
▶ Blood is carried under low pressure.
▶ Veins have less thick, muscular walls.
▶ Veins have valves at intervals to prevent the back flow of blood.
▶ Veins carry deoxygenated blood, except the pulmonary veins from the lungs.
▶ Veins are generally superficial, not deep-seated.
▶ Veins form finer blood vessels called venules which continue from capillaries.

Capillaries

▶ Capillaries are the smallest vessels.
▶ Capillaries unite arterioles and venules, forming a network in the tissues.
▶ The wall of a capillary vessel is only a single layer of cells thick. It is therefore sufficiently thin to allow the process of diffusion of dissolved substances to and from the tissues to occur.
▶ Capillaries have no valves.

▶ Blood is carried under low pressure, but higher than in veins.
▶ Capillaries are responsible for supplying the cells and tissues with nutrients.

KEY NOTE – CAPILLARY EXCHANGE

The key function of a capillary is to permit the exchange of nutrients and waste between the blood and tissue cells. Substances such as oxygen, vitamins, minerals and amino acids pass through to the tissue fluid to nourish the nearby cells, and substances such as carbon dioxide and waste are passed out of the cell. This exchange of nutrients can only occur through the semi-permeable membrane of a capillary, as the walls of arteries and veins are too thick.

Oxygenated blood flowing through the arteries appears bright red in colour due to the oxygen pigment haemoglobin; as it moves through capillaries it off loads some of its oxygen and picks up carbon dioxide. This explains why blood flow in veins appears darker.

! TASK 1

*Structural and functional differences between **arteries, veins** and **capillaries***

Complete the following table.

	Artery	Vein	Capillary
Thickness of Walls	thick, muscular walls		
Valves			none
Direction of Blood flow		towards heart	
Blood Pressure			high-low
Functional Significance	carries oxygenated blood		

The Heart

The heart is a hollow organ made up of cardiac muscle tissue which lies in the thorax above the diaphragm and between the lungs.

Composition of the Heart

It is composed of three layers of tissue:

Pericardium: the Outer Layer

This is a double-layered bag enclosing a cavity filled with pericardial fluid, which reduces friction as the heart moves during its beating.

Myocardium: the Middle Layer

This is a strong layer of cardiac muscle which makes up the bulk of the heart.

Endocardium: the Inner Layer

This lines the heart's cavities and is continuous with the lining of the blood vessels.

The heart is divided into a right and left side by a partition called a septum and each side is further divided into a thin-walled atrium above and a thick-walled ventricle below. The top chambers of the heart (the atria) take in blood from the body from the large veins and pump it to the bottom chambers. The lower chambers, the ventricles, pump blood to the body's organs and tissues.

There are four sets of valves that regulate the flow of blood though the heart:

- Between the right atrium and the right ventricle is the tricuspid valve
- Between the left atrium and the left ventricle is the bicuspid, or mistral valve.
- The aortic valve is found between the left ventricle and the aorta. The pulmonary valve is found between the pulmonary artery and the right ventricle.

The bicuspid and tricuspid valves (also known as the atrio-ventricular valves) help to maintain the direction of blood flow through the heart by allowing blood to flow into the ventricles but keeping it from returning to the atria.

The aortic and pulmonary valves are known as the semi-lunar valves; they control the blood flow out of the ventricles into the aorta and the pulmonary arteries, and prevent any backflow of blood into the ventricles. These valves open in response to pressure generated when the blood leaves the ventricles.

The heart muscle is supplied by the two coronary arteries (right and left), which originate from the base of the aorta.

KEY NOTE

If either of the coronary arteries is unable to supply sufficient blood to the heart muscle a heart attack occurs. The most common site of a heart attack is the anterior or inferior part of the left ventricle.

Blood Flow Through the Heart

Blood moves into and out of the heart in a well coordinated and precisely timed rhythm. For descriptive purposes it can be divided into the following stages:

Stage 1

Deoxygenated blood from the body enters the superior and inferior vena cava and flows into the right atrium. When the right atrium is full, it empties through the tricuspid valve into the right ventricle.

Stage 2

When the right ventricle is full, it contracts and pushes blood through the pulmonary valve into the pulmonary artery. The pulmonary artery divides into the right and left branch and takes blood to both lungs where the blood becomes oxygenated. The four pulmonary veins leave the lungs carrying oxygen rich blood back to the left atrium.

Stage 3

(This process takes place at the same time as the process described in Stage 1.)

Oxygen-rich blood leaves the left atrium and passes through the left ventricle via the bicuspid or mitral valve. When the left ventricle is full it contracts, forcing blood through the aortic valve into the aorta and to all parts of the body (except the lungs). The walls of the left ventricle are thicker in order to provide the extra strength to push blood out of the heart and around the body.

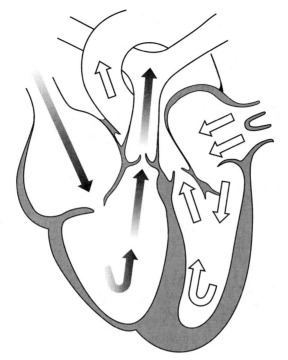

Figure 47
Blood flow through the heart

TASK 2 – THE STRUCTURE OF THE HEART

Label the following diagram of the heart:

right atrium	right ventricle	bicuspid valve
left atrium	left ventricle	tricuspid valve
septum	aortic valve	pulmonary value
arch of aorta	ascending aorta	descending aorta
superior vena cava	inferior vena cava	pulmonery artery
right pulmonary artery	right pulmonary veins	
left pulmonary artery	left pulmonary veins	

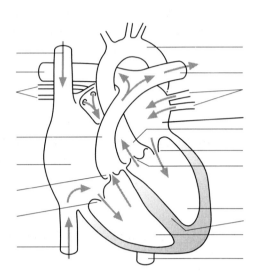

Figure 48
The structure of the heart

Function of the Heart

The function of the heart is to maintain a constant circulation of blood throughout the body. The heart acts as a pump and its action consists of a series of events known as the cardiac cycle.

The Cardiac Cycle

The cardiac cycle is the sequence of events between one heartbeat and the next, and is normally less than a second in duration.

- During a cardiac cycle, the atria contract simultaneously and force blood into the relaxed ventricles.
- The ventricles then contract very strongly and pump blood out through the aorta and the pulmonary artery.
- During ventricular contraction, the atria relax and fill up again with blood.

The heart rate can be determined by the number of cardiac cycles per minute. In an average healthy person this is likely to be between 60 and 70 cycles or beats per minute.

The heart has its own built-in rhythm. The coordinated rhythm of the heart is initiated by the built-in electrical system in the sinoatrial (SA) node, which sets the pace of the heart rate. The signal originates in the right atrium and travels to the left atrium, causing the atria to contract. At the precise moment the atria have completed their contraction, the signal travels the AV bundle to the right ventricle and into the left ventricle, causing the ventricles to contract.

KEY NOTE

If difficulty develops within the electrical system of the SA node, a device known as a pacemaker can be implanted to assist or take over initiation of the signal.

Heart Sounds

Heart sounds may be heard through a stethoscope. Closure of the heart valves produces two main sounds:

▶ The first is a low-pitched 'lubb' which is generated by the closing of the bicuspid and tricuspid valves.
▶ The second is a higher-pitched 'dubb' caused by the closing of the aortic and pulmonary valves.

Blood is transported as part of a double circuit and consists of two separate systems which are joined only at the heart.

The Pulmonary Circulation

The pulmonary circulation is the circulatory system's contribution to respiration.

This consists of the circulation of deoxygenated blood from the right ventricle of the heart to the lungs, via the pulmonary arteries. It becomes oxygenated here and is then returned to the left atrium by the pulmonary veins, to be passed to the aorta for the general or systemic circulation.

The pulmonary circulation is essentially the circulatory system between the heart and the lungs where a high concentration of blood oxygen is restored and the concentration of carbon dioxide in the blood is lowered.

The General or Systemic Circulation

The systemic circuit is the largest circulatory system and carries oxygenated blood from the left ventricle of the heart through to the aorta. Oxygenated blood is then passed around the body through the various branches of the aorta. Deoxygenated blood is returned to the right atrium via the superior and inferior vena cava, to be passed to the right ventricle to enter the pulmonary circuit.

The function of the systemic circulation is to bring nutrients and oxygen to all systems of the body and carry waste materials away from the tissues for elimination.

KEY NOTE

The increase in blood flow during a massage can help to bring fresh oxygen and nutrients into the tissues via the arterial circulation, and aid the removal of waste products via the venous circulation. Blood circulation can therefore help to improve the condition of the skin, and combined with massage, can also help to improve the muscle tone.

The Portal Circulation

Located within the systemic circuit is the portal circulation, which collects blood from the digestive organs (stomach, intestines, gall bladder, pancreas and spleen) and delivers this blood to the liver for processing, via the hepatic portal vein.

As the liver has a key function in maintaining proper concentrations of glucose, fat and protein in the blood, the hepatic portal system allows the blood from the digestive organs to take a detour through the liver to process these substances, before they enter the systemic circulation.

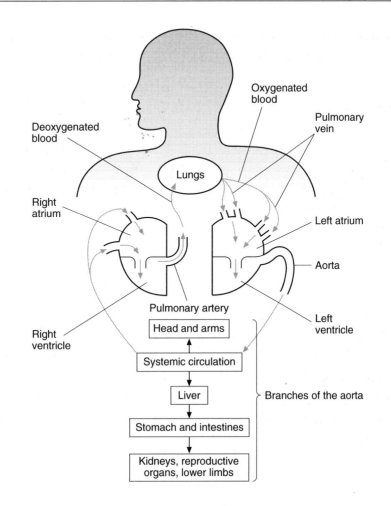

Figure 49
The pulmonary and systemic circulation

Main Arteries

The aorta is the main artery of the systemic circuit, which carries oxygenated blood around the body. It is divided into three main branches which subdivide into branches to supply the whole of the body:

1 The ascending part has branches which supply the head, neck and the top of the arms.
2 The descending thoracic part of the aorta has branches which supply organs of the thorax.
3 The descending abdominal part has branches which supply the legs and organs of the digestive, renal and reproductive systems.

The names of most major arteries are derived from the anatomical structures they serve. The femoral artery, for example, is found close to the femur. Arteries generally lie deeply-seated. They are found on both sides of the body and are identified as either right or left.

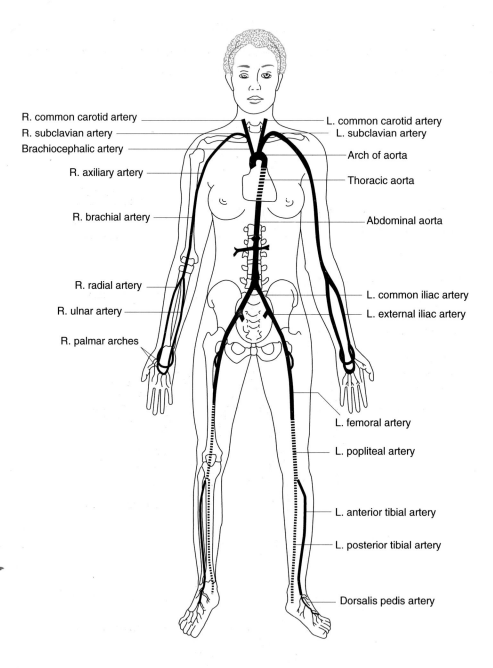

R. common carotid artery
R. subclavian artery
Brachiocephalic artery
R. axiliary artery
R. brachial artery
R. radial artery
R. ulnar artery
R. palmar arches

L. common carotid artery
L. subclavian artery
Arch of aorta
Thoracic aorta
Abdominal aorta
L. common iliac artery
L. external iliac artery
L. femoral artery
L. popliteal artery
L. anterior tibial artery
L. posterior tibial artery
Dorsalis pedis artery

Figure 50
Major systemic arteries

Main Veins

The major veins of the body are the **superior** and **inferior vena cava**, which convey deoxygenated blood from the other veins to the right atrium of the heart.

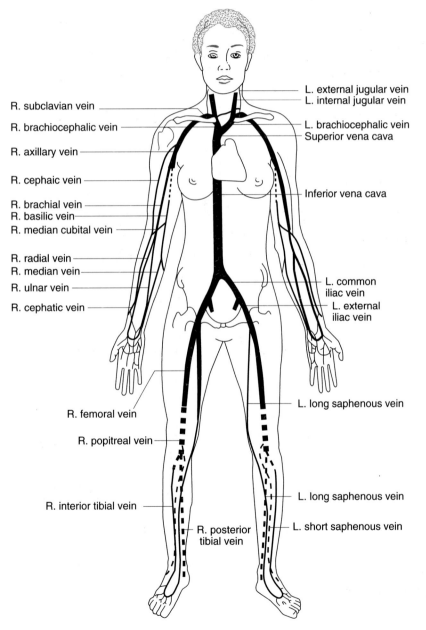

L. external jugular vein
L. internal jugular vein

R. subclavian vein

R. brachiocephalic vein

R. axillary vein

R. cephaic vein

R. brachial vein
R. basilic vein
R. median cubital vein

R. radial vein
R. median vein
R. ulnar vein

R. cephatic vein

L. brachiocephalic vein
Superior vena cava

Inferior vena cava

L. common
iliac vein
L. external
iliac vein

R. femoral vein

R. popitreal vein

R. interior tibial vein

R. posterior
tibial vein

L. long saphenous vein

L. long saphenous vein

L. short saphenous vein

Figure 51
Major systemic veins

▶ The inferior vena cava is formed by the joining of the right and left iliac veins. It receives blood from the lower parts of the body, below the diaphragm.

▶ The superior vena cava originates at the junction of the two innominate (briachiocephalic) veins. It drains blood from the upper parts of the body (head, neck, thorax and arms).

Like arteries, veins are also named for their locations and have two branches (right and left). Veins are more superficially placed than arteries.

On the previous page there is an illustration of the major systemic veins:

Arterial Blood Supply to the Head and Neck

Blood is supplied to parts within the neck, head and brain through branches of the subclavian and common carotid arteries.

Common Carotid Artery

This extends from the brachiocephalic trunk. It extends on each side of the neck and divides at the level of the larynx into two branches:

▶ The **internal carotid artery** passes through the temporal bone of the skull to supply oxygenated blood to the brain, eyes, forehead and part of the nose.

▶ The **external carotid artery** is divided into branches (facial, temporal and occipital arteries) which supply the skin and muscles of the face, side and back of the head respectively. This vessel also supplies more superficial parts and structures of the head and neck; these include the salivary glands, scalp, teeth, nose, throat, tongue and thyroid gland.

The Vertebral Arteries

These are a main division of the subclavian artery. They arise from the subclavian arteries in the base of the neck near the tip of the lungs, and pass upwards through the openings (foramina) of transverse processes of the cervical vertebrae where they unite to form a single basilar artery. The basilar artery then terminates by dividing into two posterior cerebral arteries that supply the occipital and temporal lobes of the cerebrum.

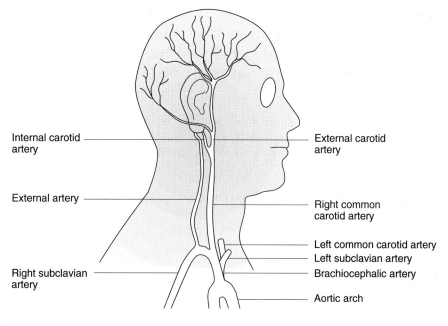

Figure 52
Blood flow to the head and neck

Venous Drainage from the Head and Neck

The majority of blood draining from the head is passed into three pairs of veins: the external jugular veins, the internal jugular veins and the vertebral veins. Within the brain, all veins lead to the internal jugular veins.

External Jugular Vein

These veins are smaller than the internal jugular veins and lie superficial to them. They receive blood from superficial regions of the face, scalp and neck. The external jugular veins descend on either side of the neck, passing over the sternomastoid muscles and beneath the platysma. They empty into the right and left subclavian veins in the base of the neck.

Internal Jugular Veins

These form the major venous drainage of the head and neck and are deep veins that parallel the common carotid artery. They collect deoxygenated blood from the brain, and pass downwards through the neck besides the common carotid arteries to join the subclavian veins.

The Vertebral Veins

These descend from the transverse openings (or foramina) of the cervical vertebrae and enter the subclavian veins. The vertebral veins drain deep structures of the neck such as the vertebrae and muscles.

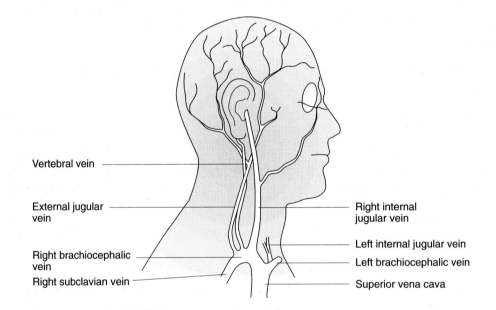

Figure 53
Venous drainage from the head and neck

KEY NOTE – BLOOD SHUNTING

Along certain circulatory pathways such as in the intestines there are strategic points where small arteries have direct connection with veins. When these connections are open, they act as shunts which allow blood in the artery to have direct access to a vein. These interconnections allow for sudden and major diversions of blood volume according to the physical needs of the body. In relation to circulation, this means that treatment should not be given after a heavy meal due to the increased circulation to the intestines, resulting in a diminished supply to other areas of the body.

Blood pressure

Blood pressure is the amount of pressure exerted by blood on an arterial wall due to the contraction of the left ventricle.

The pressure in the arteries varies during each heartbeat. The maximum pressure of the heartbeat is known as the **systolic** pressure and represents the pressure exerted on the arterial wall during active ventricular contraction. Systolic pressure can therefore be measured when the heart muscle contracts and pushes blood out into the body through the arteries.

The minimum pressure, or **diastolic** pressure, represents the static pressure against the arterial wall during rest or pause between contractions. Therefore the mimimum pressure is when the heart muscle relaxes and blood flows into the heart from the veins.

Blood pressure may be measured with the use of a sphygmomanometer.

KEY NOTE

Blood pressure is regulated by sympathetic nerves in the arterioles. An increase in stimulation of the sympathetic nervous system, as in exercise, can therefore result in a temporary increase in blood pressure.

Factors Affecting Blood Pressure

Because blood pressure is the result of the pumping of the heart in the arteries, anything that makes the heart beat faster will raise the blood pressure. Factors affecting the blood pressure include:

- excitement
- anger
- stress
- fright
- pain
- exercise
- smoking and drugs.

A normal blood pressure reading is between 100 and 140 mmHg systolic and between 60 and 90 mmHg diastolic. Blood pressure is measured in millimetres of mercury and is expressed as 120/80 mmHg.

The Pulse

The pulse is a pressure wave that can be felt in the arteries which corresponds to the beating of the heart. The pumping action of the left ventricle of the heart is so strong that it can be felt as a pulse in arteries a considerable distance from the heart. The pulse can be felt at any point where an artery lies near the surface. The radial pulse can be found by placing two or three fingers over the radial artery below the thumb. Other sites where the pulse may be felt include the carotid artery at the side of the neck and over the brachial artery at the elbow.

The average pulse in an adult is between 60 and 80 beats per minute. Factors affecting the pulse rate include:

▶ exercise
▶ heat
▶ strong emotions such as grief, fear, anger or excitement.

Disorders of the Circulatory System
Anaemia

A condition where the haemoglobin level in the blood is below normal. The main symptoms are excessive tiredness, breathlessness on exertion, pallor and poor resistance to infection. There are many causes of anaemia. It may be due to a loss of blood resulting from an accident or operation, from chronic bleeding, from iron deficiency or due to a blood disease such as leukaemia.

Aneurysm

An abnormal balloon-like swelling in the wall of an artery. This may be due to degenerative disease (congenital defects, arteriosclerosis) or any condition which causes weakening of the arterial wall such as trauma, infections, hypertension.

Angina

Pain in the left side of the chest and usually radiating to the left arm. Caused by insufficient blood to the heart muscle; usually on exertion or excitement. The pain is often described as constricting or suffocating, which can last for a few seconds or moments. Patient may become pale and sweaty. This condition indicates ischaemic heart disease.

Atheriosclerosis

A circulatory system condition characterised by a thickening, narrowing, hardening and loss of elasticity of the walls of the arteries.

Blood Pressure – High

High blood pressure is when the resting blood pressure is above normal. The World Health Organisation defines high blood pressure as consistently exceeding 160mmHg systolic and 95mmHg diastolic. High blood pressure is a common complaint and if serious may result in a stroke or a heart attack, due to the fact that the heart is made to work harder to force blood through the system. Causes of high blood pressure include:

▶ smoking
▶ obesity
▶ lack of regular exercise
▶ eating too much salt
▶ excessive alcohol consumption
▶ too much stress.

High blood pressure can be controlled by:

▶ anti-hypertensive drugs which help to regulate and lower blood pressure
▶ decreasing salt and fat intake to prevent hardening of the arteries
▶ keeping weight down
▶ giving up smoking and cutting down on alcohol consumption
▶ relaxation and leading a less stressful life.

Blood Pressure – Low

Low blood pressure is when the blood pressure is below normal and is defined by the World Health Organisation as a systolic blood pressure of 99 mmHg or less and a diastolic of less than 59 mm Hg.

Low blood pressure may be normal for some people in good health, during rest and after fatigue. The danger with low blood pressure is an insufficient supply of blood reaching the vital centres of the brain. Treatment may be by medication, if necessary.

> ### KEY NOTE
>
> High and low blood pressure are normally contra-indicated to treatments but with GP referral and an adaptation of routine, treatment may be possible. Correct positioning of the couch is essential to maximise comfort of the client with blood pressure problems and care needs to be taken to ensure that they are not lying down too long or get up too fast.

Congenital Heart Disease

A defect in the formation of the heart which usually decreases its efficiency. Defects may be in the following forms:

▶ nonclosure of the opening between the right and left ventricle – ventricular septal defects
▶ nonclosure of the opening between the right and left atrium – atrial septal defect
▶ narrowing of the aorta – coaraction of the aorta
▶ narrowing of the pulmonary artery – pulmonary stenosis
▶ nonclosure of the communication between the pulmonary artery and the aorta that exists in the foetus until delivery – patent ductus arteriosus
▶ a combination of defects.

The symptoms may vary according to the severity of the defect.

Haemophilia

A hereditary disorder in which the blood clots very slowly due to deficiency of either of two coagulation factors: Factor VIII (the antihaemophiliac factor) or Factor IX (called the Christmas factor). They are both coagulation factors normally present in blood. Deficiency of either of these factors, which are inherited by males from their mothers, results in the inability of the blood to clot (haemophilia).

The patient may experience prolonged bleeding following an injury or wound, and in severe cases there is spontaneous bleeding into the muscles and joints.

Haemophilia is controlled by a sex-linked gene which means it is almost exclusively restricted to males. Women can carry the disease, and pass it onto their sons without being affected themselves.

Heart attack (myocardial infarction)

Damage to the heart muscles which results from blockage of the coronary arteries. It can cause serious complications including heart failure.

Leukaemia

This term refers to any of a group of malignant diseases in which the bone marrow and other blood-forming organs produce an increased number of certain types of white blood cells. Overproduction of these white cells (which are immature or of abnormal form) suppresses the production of normal white cells, red cells and platelets, which leads to increased susceptibility to infection. Other manifestations or signs include enlargement of the spleen, liver and the lymph nodes, spontaneous bruising and anaemia.

Pacemaker

An artificial electrical device implanted under the skin that stimulates and controls the heart rate by sending electrical stimuli to the heart. It is usually installed for heart block and mostly placed in one side of the upper chest.

Phlebitis

Inflammation of the wall of a vein, which is most commonly seen in the legs, as a complication of varicose veins (see below). A segment of the vein becomes tender and painful, and the surrounding skin may feel hot and appear red.

Thrombosis may develop as a result of phlebitis (thrombophlebitis) with subsequent DVT (Deep Vein Thrombosis). DVT can cause clots in the lungs (or other organs) with serious consequences.

Pulmonary Embolism

A blood clot carried into the lungs, where it blocks the flow of blood to the pulmonary tissue. This is a very serious condition and can be life-threatening. Clients who suffer this condition may require hospitalisation and measures to thin the blood, for example, using warfarin. This condition presents with chest pain, cough and shortness of breath.

Raynaud's Syndrome

A disorder of the peripheral arterioles, characterised by spasm in the smooth muscle of the fingers and toes. It is generally brought on by cold or emotional upset. The effect is a pallor or discolouration of the skin due to the presence of poorly oxygenated haemoglobin.

Extremities affected can become painful and uncomfortable, and this is usually followed by redness and stiffness of the toes and fingers.

Stroke

A blocking of blood flow to the brain by an embolus in a cerebral blood vessel. A stroke can result in a sudden attack of weakness affecting one side of the body, due to the interruption to the flow of blood to the brain. A stroke can vary in severity from a passing weakness or tingling in a limb, to a profound paralysis and a coma if severe.

Sometimes the term is used to describe cerebral haemorrhage when an artery or congenital cyst of blood vessels in the brain bursts, resulting in damage to the brain and causing similar signs to thrombus of cerebral vessels. Haemorrhage is usually associated with severe headaches and can cause neck stiffness.

Thrombosis

A condition in which the blood changes from a liquid to a solid state and produces a blood clot. Thrombosis in the wall of an artery obstructs the blood flow to the tissue it supplies; in the brain this is one of the causes of a stroke, and in the heart, it results in a heart attack (coronary thrombosis). Thrombosis may also occur in a vein (deep vein thrombosis).

The thrombus (blood clot) may be detached from its site of formation and be carried in the blood to lodge in another part. See pulmonary embolism.

Varicose Veins

Veins are known as varicose when the valves within them lose their strength. As a result of this, blood flow may become reversed or static. Valves are concerned with preventing the back flow of blood, but when their function is impaired they are unable to prevent the blood from flowing downwards, hence the walls of the affected veins swell and bulge out and become visible through the skin. Varicose veins may be due to several factors:

▶ hereditary tendencies
▶ ageing
▶ obesity, as excess weight puts pressure on the walls of the veins
▶ pregnancy
▶ sitting or standing motionless for long periods of time, causing pressure to build up in the vein.

KEY NOTE

Varicose veins can be extremely painful and great care needs to be taken with a client with varicose veins. Treatment is therefore contra-indicated in the area affecting the veins.

SELF-ASSESSMENT QUESTIONS – THE CIRCULATORY SYSTEM

1. State the composition of blood.

2. State the functional significance of the following blood cells

a) erythrocytes

b) leucocytes

c) thrombocytes

3. List the main functions of blood.

4. What is the function of an artery in blood transport?

5. What is the function of a vein in blood transport?

6. Why is a capillary one cell layer thick?

7. Briefly describe the position and structure of the heart.

8. Briefly describe the flow of blood through the heart.

9. Briefly outline the cardiac cycle

10. Explain what is meant by the following:

a) the pulmonary circulation

b) the systemic circulation

11. Identify the following blood vessels from the description given

a) main artery of the systemic circuit. Carries oxygenated blood around the body through its three branches

b) two great arteries of the neck that supply oxygenated blood to the head and neck

c) these vessels have an internal and external branch. They carry deoxygenated blood from the head and neck.

d) this blood vessel has a right and left branch and carries deoxygenated blood from the heart to the lungs.

e) these vessels have an internal and external branch. They carry deoxygenated blood from the head and neck.

f) This vessel is a main vein that carries deoxygenated blood from the upper part of the body back to the heart.

g) This vessel carries deoxygenated blood from the lower part of the body back to the heart.

h) There are four of these vessels that carry oxygenated blood from the lungs back to the heart.

12. Explain what is meant by the following terms:

a) blood pressure

b) pulse

13. List four factors which may increase blood pressure.

14. State two factors that may cause the pulse rate to rise.

15. State how the following may affect the provision of therapeutic treatments:

a) high blood pressure

b) blood shunting

c) varicose veins

CHAPTER 9

The Lymphatic System

The lymphatic system is a one-way drainage system for the tissues. It helps to provide a circulatory pathway for tissue fluid to be transported, as lymph, from the tissue spaces of the body into the venous system, where it becomes part of the blood circulation. Through the filtering action of the lymph nodes, along with specific organs such as the spleen, the lymphatic system also helps to provide immunity against disease.

A competent therapist needs to be able to:

▶ understand the connection between blood and lymph in order to carry out treatments effectively

By the end of this chapter you will be able to relate the following knowledge to your practical work carried out in the salon:

▶ what lymph is and how it is formed
▶ the connection between blood and lymph
▶ the circulatory pathway of lymph
▶ the functions of the lymphatic system
▶ the names and positions of the main lymph nodes of the head, neck and the body
▶ the drainage of lymph from the head, neck and the body
▶ disorders of the lymphatic system.

What is Lymph?

Lymph is a transparent, colourless, watery fluid which is derived from tissue fluid and is contained within lymph vessels. It resembles blood plasma in composition, except that it has a lower concentration of plasma proteins. This is because some large protein molecules are unable to filter through the cells forming the capillary walls so they remain in blood plasma. Lymph contains only one type of cell; these are called lymphocytes.

How is Lymph Formed?

As blood is distributed to the tissues, some of the plasma escapes from the capillaries and flows around the tissue cells, delivering nutrients such as oxygen and water to the cell and picking up cellular waste such as urea and carbon dioxide. Once the plasma is outside the capillary and is bathing the tissue cells, it becomes tissue fluid.

Some of the tissue fluid passes back into the capillary walls to return to the blood stream via the veins, and some is collected up by a lymph vessel where it becomes lymph. Lymph is then taken through its circulatory pathway and is ultimately returned to the bloodstream.

The Connection between Blood and Lymph

The lymphatic system is therefore often referred to as a secondary circulatory system as it consists of a network of vessels that assist the blood in returning fluid from the tissues back to the heart.

In this way, the lymphatic system is a complementary system for the circulatory system. After draining the tissues of excess fluid, the lymphatic system returns this fluid to the cardiovascular system. This helps to maintain blood volume, blood pressure and prevent oedema.

Functions of the Lymphatic System

▶ The lymphatic system is important for the distribution of fluid and nutrients in the body, because it drains excess fluid from the tissues and returns to the blood protein molecules, which are unable to pass back through the blood capillary walls because of their size.
▶ The lymph nodes help to fight infection by filtering lymph and destroying invading micro-organisms. Lymphocytes are reproduced in the lymph node and following infection they generate antibodies to protect the body against subsequent infection. Therefore the lymphatic system plays an important part in the body's immune system.

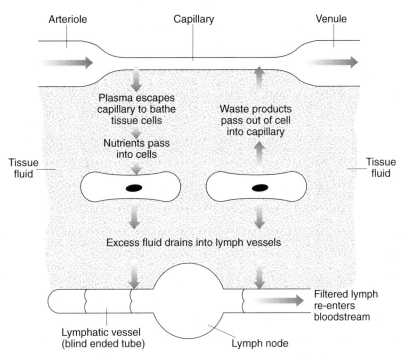

Figure 54

The connection between blood and lymph

▶ The lymphatic system also plays an important part in absorbing the products of fat digestion from the villi of the small intestine. While the products of carbohydrate and protein digestion pass directly into the bloodstream, fats pass directly into the intestinal lymph vessels, known as *lacteals*.

Structure of the Lymphatic System

The lymphatic system contains the following structures:

▶ lymph capillaries
▶ lymph vessels
▶ lymph nodes
▶ lymph collecting ducts.

Lymph Capillaries

Lymph capillaries commence in the tissue spaces of the body as minute blind-end tubes, as lymph is a one-way circulatory pathway. The walls of the lymph capillaries are like those of the blood capillaries in that they are a single cell layer thick to make it possible for tissue fluid to enter them. However, they are permeable to substances of larger molecular size than those of the blood capillaries.

The lymph capillaries mirror the blood capillaries and form a network in the tissues, draining away excess fluid and waste products from the tissue spaces of the body. Once the tissue fluid enters a lymph capillary it becomes lymph and is gathered up into larger lymph vessels.

KEY NOTE – OEDEMA

The term oedema refers to an excess of fluid within the tissue spaces that causes the tissues to become waterlogged.

Lymph Vessels

Lymph vessels are similar to veins in that they have thin collapsible walls and their role is to transport lymph through its circulatory pathway. They have a considerable number of valves which help to keep the lymph flowing in the right direction and prevent backflow. Superficial lymph vessels tend to follow the course of veins by draining the skin, whereas the deeper lymph vessels tend to follow the course of arteries and drain the internal structures of the body.

Networks, or plexuses of lymph channels exist throughout the body. These intertwined channels are found in the following areas:

▶ **Mammary** plexus: lymph vessels around the breasts
▶ **Palmar** plexus: lymph vessels in the palm of the hand
▶ **Plantar** plexus: lymph vessels in the sole of the foot.

The lymph vessels carry the lymph towards the heart under steady pressure and about two to four litres of lymph pass into the venous system every day. Once lymph has passed through the lymph vessels it drains into at least one lymphatic node before returning to the blood circulatory system.

KEY NOTE

As the lymphatic system lacks a pump, lymphatic vessels have to make use of contracting muscles that assist the movement of lymph. Therefore, lymphatic flow is at its greatest during exercise due to the increased contraction of muscle.

Lymph Nodes

A lymph node is an oval or bean shaped structure, covered by a capsule of connective tissue. It is made up of lymphatic tissue and is divided into two regions: an outer cortex and an inner medulla.

There are more than 100 lymph nodes, placed strategically along the course of lymph vessels. They vary in size between one to 25 millimetres in length and are massed in groups; some are superficial and lie just under the skin, whereas others are deeply seated and are found near arteries and veins.

Each lymph node receives lymph from several afferent lymph vessels, and blood from small arterioles and capillaries. Valves of the afferent lymph vessels open towards the node, therefore lymph in these vessels can only move towards the mode. Lymph flows slowly through the node, moving from the cortex to the medulla, and leaves through an efferent vessel which opens away from the node.

▶ The **afferent** vessels enter a lymph node.
▶ The **efferent** vessels drain lymph from a node.

The function of a lymph node is to act as a filter of lymph to remove or trap any micro-organisms, cell debris, or harmful substances which may cause infection, so that, when lymph enters the blood, it has been cleared of any foreign matter.

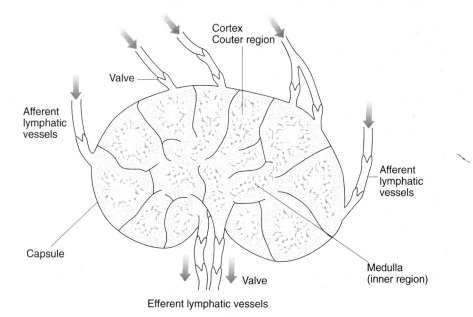

Figure 55
The structure of a lymph node

When lymph enters a node, it comes into contact with two specialised types of leucocytes:

▶ **monocytes**, which are phagocytic in action; they engulf and destroy dead cells, bacteria and foreign material in the lymph
▶ **lymphocytes**, which are reproduced within the lymph nodes and can neutralise invading bacteria, and produce chemicals and antibodies to help fight disease.

KEY NOTE

If an area of the body becomes inflamed or otherwise diseased, the nearby lymph nodes will swell up and become tender, indicating that they are actively fighting the infection.

Once filtered, the lymph leaves the node by one or two efferent vessels, which open away from the node. Lymph nodes occur in chains, so that the efferent vessel of one node becomes the afferent vessel of the next node, in the pathway of lymph flow. Lymph drains through at least one lymph node before it passes into two main collecting ducts before it is returned to the blood.

Lymphatic Ducts

From each chain of lymph nodes, the efferent lymph vessels combine to form lymph trunks which empty into two main ducts: the **thoracic** and the **right lymphatic** ducts. These ducts collect lymph from the whole body and return it to the blood via the subclavian veins.

▶ The **thoracic duct:** this is the main collecting duct of the lymphatic system. It is the largest lymph vessel in the body and extends from the second lumbar vertebra up through the thorax to the root of the neck. The thoracic duct collects lymph from the left side of the head and neck, the left arm, the lower limbs and abdomen and drains into the left subclavian vein to return it to the bloodstream
▶ The **right lymphatic duct:** this duct is very short in length. It lies in the root of the neck and collects lymph from the right side of the head and neck and the right arm, and drains into the right subclavian vein, to be returned to the bloodstream.

Lymphatic Drainage

Movement of lymph throughout the lymphatic system is known as **lymphatic**

drainage, and it begins in the lymph capillaries. The movement of lymph out of the tissue spaces and into the lymphatic capillaries is assisted by:

▶ the pressure exerted by the skeletal muscles against the vessels during movement
▶ changes in internal pressure during respiration
▶ compression of lymph vessels from the pull of the skin and fascia during movement.

KEY NOTE

Factors such as muscle tension put pressure on the lymphatic vessels and may block them, interfering with efficient drainage.

As the muscles relax during a massage for instance, the lymphatic vessels open to stimulate the flow of lymph thereby assisting drainage.

The client can help the process by taking slow deep breaths which help to stimulate the lymph flow.

Lymphatic Drainage of the Head and Neck

The main groups of lymph nodes relating to the head and neck:

Name of Lymph Nodes	Position	Areas Lymph is Drained From
Cervical Nodes (Deep)	deep within the neck, located along the path of the larger blood vessels (carotid artery and internal jugular vein)	drain lymph from the larynx, oesophagus, posterior of the scalp and neck, superficial part of chest and arm
Cervical Nodes (Superficial)	located at the side of the neck, over the sternomastoid muscle	drain lymph from the lower part of the ear and the cheek region
Submandibular Nodes	beneath the mandible	drain chin, lips, nose, cheeks and tongue

Name of Lymph Nodes	Position	Areas Lymph is Drained From
Occipital Nodes	at the base of the skull	drains back of scalp and the upper part of the neck
Mastoid Nodes (Postauricular)	behind the ear in the region of the mastoid process	drains the skin of the ear and the temporal region of the scalp
Parotid Nodes	at the angle of the jaw	drains nose, eyelids and ear

TASK 1

Label the following lympth nodes relating to the head end neck on Figure 56:

superficial cervical nodes   deep cervical nodes   mastoid nodes

parotid nodes   occipital nodes   submandibular nodes

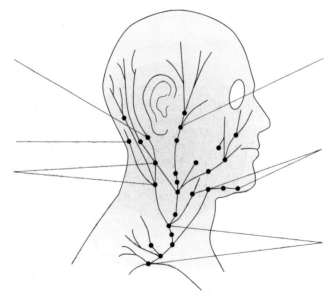

Figure 56
Major lymph of the head and neck

Lymphatic Drainage of the Body

Lymph nodes are mainly clustered at joints, where they assist in pumping lymph through the nodes when the joint moves. The superficial lymph nodes are most numerous in the groin, axillae and the neck; most of the deep lymph nodes are found alongside blood vessels of the pelvic, abdominal and thoracic cavities.

The main groups of lymph nodes relating to the body are as follows:

Name of Lymph Nodes	Position	Areas Lymph is Drained From
Cervical Nodes (Deep)	deep within the neck, located along the path of the larger blood vessels	drain lymph from the larynx, oesophagus, posterior of the scalp and neck, superficial part of chest and arm
Cervical Nodes (Superficial)	located at the side of the neck, over the sternomastoid muscle	drain lymph from the lower part of the ear and the cheek region
Axillary Nodes	in the underarm region	drain the upper limbs, wall of the thorax, breasts, upper wall of the abdomen
Supratrochlear / Cubital Nodes	in the elbow region (medial side)	upper limbs, which passes through the axillary nodes
Thoracic Nodes	within the thoracic cavity and along the trachea and bronchi	organs of the thoracic cavity and from the internal wall of the thorax
Abdominal Nodes	within the abdominal cavity, along the branches of the abdominal aorta	organs within the abdominal cavity
Pelvic Nodes	within the pelvic cavity, along the paths of the iliac blood vessels	organs within the pelvic cavity

Name of Lymph Nodes	Position	Areas Lymph is Drained From
Inginual	in the groin	lower limbs, the external genitalia and the lower abdominal wall
Popliteal	behind the knee	the lower limbs through deep and superficial nodes

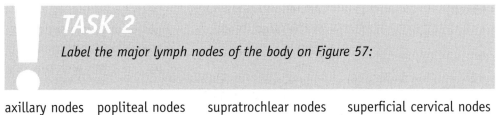

TASK 2

Label the major lymph nodes of the body on Figure 57:

axillary nodes   popliteal nodes   supratrochlear nodes   superficial cervical nodes

pelvic nodes   thoracic nodes   abdominal nodes   inginual nodes

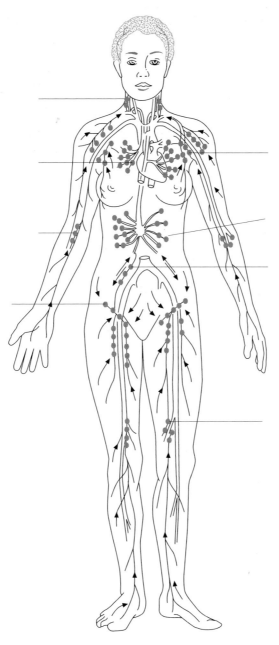

Figure 57
Major lymph nodes of the body

Summary of the Circulatory Pathway of Lymph

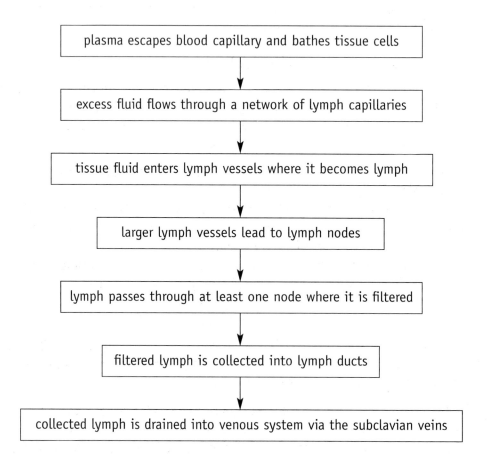

Lymphatic Organs

Lymphatic organs, whose functions are closely related to those of the lymph nodes, are the spleen, the tonsils and the thymus.

The Spleen

The spleen is the largest of the lymphatic organs, and is located in left hand side of the abdominal cavity between the diaphragm and the stomach. As the spleen is largely a mass of lymphatic tissue, it contains lymph nodes which produce lymphocytes and monocytes, which are phagocytic.

- The spleen is a major site for filtering out worn out red blood cells and destroying micro-organisms in the blood.
- The spleen is concerned with protection from disease and the manufacture of antibodies; it functions with the lymphatic system by storing lymphocytes and releasing them as part of the immune response.
- The spleen also serves as a blood reservoir and can release small amounts of blood into the circulation during times of emergency or blood loss.

The Tonsils

The tonsils are composed of lymphatic tissue and are located in the oral cavity and the pharynx. There are three different sets of tonsils, all of which provide defence against micro-organisms that enter the mouth and nose.

- The palatine tonsils are the set commonly identified as the tonsils, and are located at the back of the throat, one on each side.
- The pharyngeal tonsils are known as the adenoids, and they lie on the wall of the nasal part of the pharynx.
- The third set, the lingual tonsils, are found below the tongue.

The Thymus

The thymus gland is a triangular-shaped gland composed of lymphatic tissue. It is located in the upper chest, above the superior vena cava and below the thyroid, where it lies against the trachea. The function of the thymus is important in the new-born baby, in promoting the development and maturation of certain lymphocytes and in programming them to become T-cells of the immune system.

The thymus gland begins to atrophy after puberty and becomes only a small remnant of lymphatic tissue in adulthood.

Disorders of the Lymphatic System
Hodgkin's Disease

Malignant disease of the lymphatic tissues, usually characterised by painless enlargement of one or more groups of lymph nodes in the neck, armpit, groin, chest, or abdomen. The spleen, liver, bone marrow and bones may also be involved. Apart from the enlarging nodes there may also be weight loss, fever, profuse sweating at night and itching.

Oedema

An abnormal swelling of body tissues due to an accummulation of tissue fluid. It could be the result of heart failure, liver or kidney disease or due to chronic varicose veins. The resultant swelling of the tissues may be localised, as with an injury or inflammation, or may be more generalised, as in heart or kidney failure.

Subcutaneous oedema commonly occurs in the legs and ankles due to the influence of gravity, and is a common problem in women before menstruation and in the last trimester of pregnancy.

SELF ASSESSMENT QUESTIONS – THE LYMPHATIC SYSTEM

1. What is lymph?

2. What is the composition of lymph?

3. State the connection between blood and lymph.

4. Describe the structure of a lymph vessel.

....

....

....

....

5. Briefly describe the structure of a lymph node. Draw a simplified diagram to illustrate your answer.

....

....

....

....

....

6. Complete the following in relation to the circulatory pathway of lymph:

a) lymph capillaries

....

b) lymph

....

c) lymph

....

d) lymphatic ducts

....

e) filtered lymph returns to bloodstream via

....

7. Identify the following:

a) the nodes found at the base of the skull

....

b) the nodes that are located at the side of the neck over the sternomastoid muscle

....

c) the nodes that are located beneath the mandible

....

d) the nodes that are located deep within the neck, along the path of the carotid artery and internal jugular vein

....

8. State the areas of the body the following lymph nodes drain lymph from:

a) axillary nodes

....

b) inginual nodes

....

c) popliteal nodes

....

d) supratrochlear/ cubital nodes

....

9. Name the two main lymphatic ducts and the areas of the body they collect lymph from.

....

....

10. State two functions of the lymphatic system.

....

....

....

11. Briefly state the structure and functions of the spleen.

....

....

....

12. Name two other forms of lymphatic tissue, other than the spleen.

....

....

CHAPTER 10

The Immune System

The human body is equipped with a variety of defence mechanisms that prevent the entry of foreign agents, known as pathogens. This defence is called immunity. When working effectively, the immune system protects the body from most infectious micro-organisms. It does this both directly by cells attacking the micro-organisms, and indirectly by releasing chemicals and protective antibodies.

It is useful for a therapist to have a basic knowledge of immunity in order to be able to understand the body both in health and in disease.

By the end of this chapter you will be able to recall and understand the following:

▶ what is meant by the term 'immunity'
▶ the different types of immunity – specific and non-specific
▶ the immune response
▶ what is meant by the term 'allergy'
▶ disorders of the immunity system.

What is the Immune System?

The immune system is not a specific structural organ system, but more of a functional system. It draws upon the structures and processes of each of the organs, tissues and cells of the body, and the chemicals produced in them, to eliminate any pathogen, foreign substance or toxic material that can be damaging to the body.

Immunity can therefore be defined as:

▶ the ability of the body to resist infection and disease by the activiation of specific defence mechanisms.

The human body has a variety of different defence mechanisms; some are *non-specific* in that they do not differentiate between one threat and another. Others are *specific*, as the body mounts its defence specifically against a particular kind of threat.

Non-specific Immunity

Non-specific immunity is programmed genetically in the human body from birth. The non-specific defences that are present from birth include:

▶ mechanical barriers
▶ chemicals
▶ inflammation
▶ phagocytosis
▶ fever

Mechanical Barriers

These are barriers such as the skin and mucous membrane that line the tubes of the respiratory, digestive, urinary and reproductive systems. As long as these barriers remain unbroken, many pathogens are unable to penetrate them.

▶ The respiratory system is lined with mucous-secreting cells to help remove micro-organisms from the respiratory tract.
▶ The highly acidic environment in the stomach can help to kill pathogens, along with the production of saliva, which has an antimicrobial effect.
▶ Urine helps to deter the growth of micro-organisms in the genito-urinary tract.
▶ The ph of the vagina protects against the multiplication and growth of microbes.

Chemicals

Chemicals are liberated by different cells that play an important role in immunity. There are many different types of chemicals involved in immunity including interferons, complements and histamine.

Interferons

These are proteins produced by cells infected by viruses. Interferon forms antiviral proteins to help protect uninfected cells and inhibit viral growth. There are three types of human interferon:

1 alpha (from white blood cells)
2 beta (from fibroblasts)
3 gamma (from lymphocytes)

Complements

These are proteins found in blood that combine to create substances that phagocytise (ingest) bacteria.

Histamine

This is a chemical released by a variety of tissue cells. This includes mast cells, basophils (a type of white blood cell), and platelets. The release of histamine causes vasodilation to bring more blood to the area of injury or infection. It also increases vascular permeability, to allow fluid to enter the damaged area and dilute the toxins released.

Inflammation

Inflammation is a sequence of events involving chemical and cellular activation that destroys pathogens and aids in the repair of tissues. It is a tissue response and symptoms include localised redness, swelling, heat and pain.

The major actions that occur during an inflammation response include the following:

▶ The blood vessels dilate, resulting in an increase in blood volume (hyperaemia) to the affected area.
▶ Capillary permeability increases, causing tissues to become red, swollen, warm and painful.
▶ White blood cells invade the area, and help to control pathogens by phagocytosis.
▶ In the case of bacterial infections, pus may form.
▶ Body fluids collect in the inflamed tissues. These fluids contain fibrinogen and other blood factors that promote clotting.
▶ Fibroblasts may appear and a connective tissue sac may be formed around the injured tissues.
▶ Phagocytic cells remove dead cells and other debris from the site of inflammation.
▶ New cells are formed by cellular reproduction to replace dead injured ones.

Phagocytosis

Neutrophils and **monocytes** are the most active phagocytic cells of the blood. Neutrophils are able to engulf and ingest smaller particles, while monocytes can phagocytise larger ones. Moncoytes give rise to macrophages (large scavenger cells), which become fixed in various tissues and attached at the inner walls of the blood and lymphatic vessels.

Fever

An individual is said to have a fever if their body temperature is maintained above 37.2°C (99°F). The increase in temperature during a fever tends to inhibit some viruses and bacteria. It also speeds up the body's metabolism and thereby increases the activity of defence cells.

Specific Immunity

Immunity involves interaction between two types of molecule – an antigen and an antibody.

▶ An **antigen** is any substance that the body regards as foreign or potentially dangerous, and against which it produces an antibody.
▶ An **antibody** is a specific protein produced to destroy or suppress antigens.

Antibodies circulate in the blood and tissue fluid, killing germs or making them harmless. Antibodies also neutralise poisonous chemicals called toxins which germs produce.

Specific immunity involves very specific responses to each identified foreign substance and calls on special memory cells to help if the invader reappears. It is the ability to recognise certain antigens and destroy them. The body must be able to identify which substances are capable of causing a threat before any type of response can be initiated.

KEY NOTE

There are many different types of antigens, and each requires a specific antibody to destroy it. Antibodies are manufactured in such a way that they are specific to antigens. For example, if a person has been exposed to the chicken pox virus, plasma cells manufacture antibodies specific to the chicken pox virus.

Memory cells are also formed against the virus. If the person is exposed a second time to the same virus, then memory cells are stimulated and large quantities of antibodies against the chicken pox virus are produced.

How Antibodies Work

Antibodies work in many different ways; some neutralise the antigens when they combine with them and prevent them from carrying out their effects. Others may lyse (destroy) the cell on which the antigen is present. When antibodies are bound to antigens on the surface of bacteria, they attract other white blood cells like **macrophages** to engulf them.

The key cells of specific immunity are a specialised group of white blood cells called **lymphocytes**. They are capable not only of recognising foreign agents, but remembering the agents they have encountered, and therefore are able to react more rapidly and with greater force if they encounter the agent again.

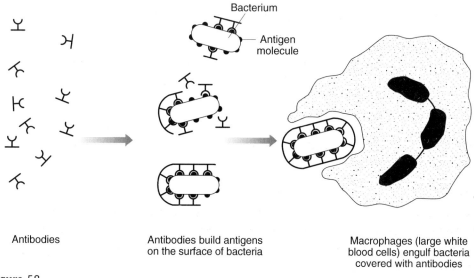

Antibodies

Antibodies build antigens on the surface of bacteria

Macrophages (large white blood cells) engulf bacteria covered with antibodies

Figure 58
How antibodies destroy germs

The Immune Response

There are two types of immune response produced by different types of lymphocytes:

1 **Humoral immunity** – this involves the B-lymphocytes which produce free antibodies that circulate in the bloodstream.
2 **Cell-mediated immunity** – this is effected by the helper T-cells, suppressor T-cells and natural killer (NK) cells that recognise and respond to certain antigens to protect the body against their effects.

Lymphocytes develop in the following three ways:

 T-cells begin in the bone marrow and grow in the thymus gland. They are able to

recognise antigens and respond by releasing inflammatory and toxic materials. Specialised T-cells also regulate the immune response, either by amplifying the response (T4 cells) or by suppressing the body's response (T8 cells). Some T-cells develop into memory cells and handle secondary response on re-exposure to antigens that have already produced a primary response.

▶ B-cells grow and develop in the bone marrow. B-cells contain immunoglobulin, an antibody that responds to specific antigens. Some B-cells modify and become non-antigen specific, which means that they have a greater ability to respond to bacterial and viral pathogens. Some B-cells become memory cells and are able to deal with re-exposure to antigens.

▶ A type of lymphocyte that does not develop the same structural or functional characteristics as the T-cells or B-cells are the natural killer cells (NK) cells. They also develop in the bone marrow, and when mature can attack and kill tumour cells and virus-infected cells during their initial developmental stage, before the immune system is activated.

KEY NOTE

Macrophages, along with the T-cells and B-cells, make up the body's cellular immunity system, and are found in high concentrations in the thymus and the spleen. Their role is to constantly monitor the contents of body fluid as it flows through the lymphatic system, where lymph nodes filter it. If foreign matter is detected, mature lymphocytes concentrate in large numbers in the lymph nodes, and migrate to the bloodstream to deal with it. This concentration causes the lymph nodes in the neck, armpits or groin to swell during some illnesses.

Primary and Secondary Responses

The initial response of the body on first exposure to antigens is know as the **primary** response. It normally takes about two weeks after exposure to the antigen for antibody levels to peak. This is due to the fact that B-cells have to become converted to plasma cells that secrete antibodies specifically against the antigen.

If the individual is exposed to the antigen the second time, the presence of memory cells stimulates rapid production of antibodies and this is known as the **secondary** response. The antibody levels are much higher than the primary response and remain elevated for a very long time. Secondary response can occur even if many years have elapsed since the first exposure to the antigen.

Immunisation

The body may be artificially stimulated into producing antibodies, and this is known as **immunisation**. This prepares the body to ward off infection in advance, and is carried out by inoculating an individual with a vaccine (a liquid containing antigens powerful enough to stimulate antibody formation without causing harm). Vaccines have been developed against many diseases including Diphtheria, Polio, Tetanus, Whooping Cough and Measles.

Allergy

Under certain circumstances, abnormal responses or allergic reactions may occur when a foreign substance, or antigen, enters the body. An allergic reaction can only occur if the person has already been exposed to the antigen at least once before and has developed some antibody to it.

The type and severity of an allergic reaction depends upon the strength and persistence of the antibody screen evoked by previous exposure to the antigen. These antibodies are located on the cells in the skin or mucous membranes of the respiratory and gastro-intestinal tracts. Typical antigens include pollen, dust, feathers, wool, fur, certain foods and drugs.

The reactions may cause symptoms of hay fever, asthma, eczema, urticaria and contact dermatitis. If there is much cellular damage, excessive amounts of histamine may be released causing circulatory failure (anaphylaxis).

KEY NOTE

One of the effects of stress is an increase in the levels of glucocorticoid steroids. These depress the inflammatory response by inhibiting the mast cells and reducing the number and activity of phagocytic cells in the tissues.

During times of stress, or severe depression, the T-lymphocytes become depressed, which weaken the immune system and increase susceptibility to illness. A person with a depressed immunity will usually complain of recurrent infections, as the body is unable to mount a proper immune response to the invading pathogens.

Disorders of the Immune System

AIDS

This is a viral infection which progressively destroys the immunity of the individual. It is caused by the Human Immunodeficiency Virus (HIV), which primarily affects the T-lymphocytes, resulting in the suppression of the body's immune response.

AIDS patients become vulnerable to infections that do not affect normal individuals; infections that produce mild symptoms otherwise may produce severe symptoms in them. Patients may also be prone to usual cancers.

This syndrome is caused by contact with infected blood or body fluids. It is common in drug addicts using infected injection needles and syringes, through having unprotected sexual intercourse and is noticeabley high in homosexuals. It is important to remember that haemophiliacs are prone to have AIDS due to infected blood from transfusions.

Lupus Erythematosus

This is a chronic inflammatory disease of connective tissue affecting the skin and various internal organs. It is an auto-immune disease and can be diagnosed by the presence of abnormal antibodies in the bloodstream.

Typical signs are a red scaly rash on the face, arthritis and progressive damage to the kidneys. Often the heart, lungs and brain are also affected by progressive attacks of inflammation, followed by the formation of scar tissue. It can also cause psychiatric illness due to direct brain involvement, but in a milder form, only the skin is affected.

Myalgic Encephalomyelitis (chronic fatigue syndrome)

A condition which is characterised by extreme disabling fatigue that has lasted for at least six months, is made worse by physical or mental exertion and is not resolved by bed rest. The symptom of fatigue is often accompanied by some of the following:

- muscle pain or weakness
- poor coordination
- joint pain
- slight fever
- sore throat
- painful lymph nodes in the neck and armpits
- depression
- inability to concentrate
- general malaise.

It is thought to be associated with a reaction to a viral infection in those with an abnormal immune response. This abnormal response is influenced by age, genetic predisposition, gender, stress, environment and previous illness. It can happen in any age group, but recently children and adolescents are noticed to have a higher incidence.

SELF ASSESSMENT QUESTIONS – THE IMMUNE SYSTEM

1. What is meant by the term 'immunity'?

2. What is the difference between specific immunity and non-specific immunity?

3. Define the terms antigen and antibody.

4. The key cells of specific immunity are called

5. Explain what is meant by the term 'allergy'.

6. Briefly describe the effects of stress on the immune system.

CHAPTER 11

The Respiratory System

The respiratory system consists of the nose, the naso-pharynx, the pharynx, the larynx, the trachea, the bronchi and the lungs which provide the passageway for air, in and out of the body. The mechanism of respiration is the process by which the living cells of the body receive a constant supply of oxygen and remove carbon dioxide.

Oxygen is needed by every cell of the body for survival and delivery; respiration is the process by which the living cells of the body receive a constant supply of oxygen, and remove carbon dioxide and other gases.

Our respiratory system serves us in many ways, in that it exchanges oxygen and carbon dioxide, detects smell, produces speech and regulates ph.

A competent therapist needs to have knowledge of the mechanism of respiration in order to understand the importance of correct breathing.

By the end of this chapter you will be able to relate the following knowledge to your practical work carried out in the salon:

▶ the functional significance of the main structures of the respiratory system
▶ the process of the interchange of gases in the lungs
▶ the mechanism of breathing
▶ the importance of correct breathing
▶ disorders of the respiratory system.

The Structures of the Respiratory System

The Nose

The nose is divided into the right and left cavities. It is lined with tiny hairs called cilia (which begin to filter the incoming air) and mucous membrane (which secretes a sticky fluid called mucus to prevent dust and bacteria from entering the lungs). The nose moistens, warms and filters the air and is an organ which senses smell.

The Naso-Pharynx

The naso-pharynx is the upper part of the nasal cavity behind the nose, and is lined with mucous membrane. The naso-pharynx continues to filter, warm and moisten the incoming air.

The Pharynx

The pharynx or throat is a large cavity which lies behind the mouth and between the nasal cavity and the larynx. The pharynx serves as an air and food passage but cannot be used for both purposes at the same time, otherwise choking would result. The air is also warmed and moistened further as it passes through the pharynx.

The Larynx

The larynx is a short passage connecting the pharynx to the trachea, and contains the vocal cords. The larynx has a rigid wall and is composed mainly of muscle and cartilage, which help to prevent collapse and obstruction of the airway. The larynx provides a passageway for air between the pharynx and the trachea.

The Trachea

The trachea or windpipe is made up mainly of cartilage, which helps to keep the trachea permanently open. The trachea passes down into the thorax and connects the larynx with the bronchi, which pass into the lungs.

The Bronchi

The bronchi are two short tubes, similar in structure to the trachea, which lead to and carry air into each lung. They are lined with mucous membrane and ciliated cells and, like the trachea, contain cartilage to hold them open. The mucus traps solid particles and cilia move them upwards, preventing dirt from entering the delicate lung tissue. The bronchi subdivide into **bronchioles** in the lungs. These subdivide yet again and finally end in minute air-filled sacs called **alveoli**.

The Lungs

The lungs are cone-shaped spongy organs, situated in the thoracic cavity on either side of the heart. Internally the lungs consist of tiny air sacs called alveoli, which are arranged in lobules and resemble bunches of grapes. The function of the lungs is to facilitate the exchange of the gases oxygen and carbon dioxide. In order to carry this out efficiently, the lungs have several important features:

▶ a very large surface area provided by the alveoli
▶ thin permeable membrane surrounding the walls of the alveoli
▶ a thin film of water lining the alveoli, which is essential for dissolving oxygen from the alveoli air
▶ thin-walled blood capillaries forming a network around the alveoli, which absorb oxygen from the air breathed into the lungs and release carbon dioxide into the air breathed out of the alveoli.

The structures enclosed within the lungs are bound together by elastic and connective tissue. On the outside, the lungs have a serous covering or membrane called a **pleura**, which prevents friction between the lungs and the chest wall.

TASK 1 – STRUCTURES OF THE RESPIRATORY SYSTEM

Label the structures of the respiratory system on Figure 59:

naso-pharynx	larynx	bronchus	pharynx	trachea	bronchioles
intercostal muscle	rib	diaphragm	pleura	alveoli	

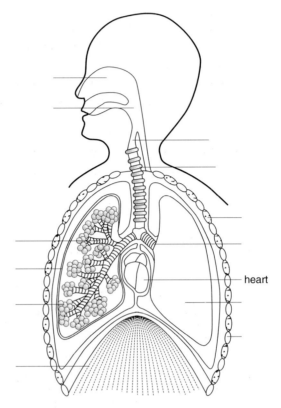

heart

Figure 59
Structures of the respiratory system

Oxygen and carbon dioxide exchange is the primary function of the respiratory system. Oxygen is needed by every cell of the body, and delivery is accomplished by way of the bloodstream. The respiratory and circulatory systems therefore both participate in the respiratory process.

The Interchange of Gases in the Lungs

The interchange of gases in the lungs involves the absorption of oxygen from the air in exchange for carbon dioxide, which is released by the body as a waste product of cell metabolism:

▶ During inhalation, oxygen is taken in through the nose and mouth. It flows along the trachea and bronchial tubes to the alveoli of the lungs, where it diffuses through the thin film of moisture lining the alveoli.
▶ The inspired air, which is now rich with oxygen, comes into contact with the blood in the capillary network surrounding the alveoli.
▶ The oxygen then diffuses across a permeable membrane wall surrounding the

alveoli, to be taken up by red blood cells. The oxygen-rich blood is carried to the heart, and is then pumped to cells throughout the body.

▶ Carbon dioxide, collected by the respiring cells around the body, passes in the opposite direction by diffusing from the capillary walls into the alveoli; it is passed through the bronchi and trachea, and exhaled through the nose and mouth.

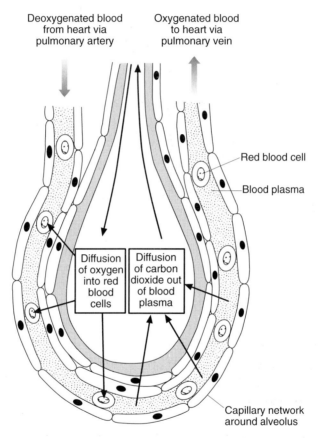

Figure 60
The interchange of gases

The Mechanism of Respiration

The mechanism of respiration is the means by which air is drawn in and out of the lungs. It is an active process where the muscles of respiration contract to increase the volume of the thoracic cavity. Air is moved in and out of the lungs by the combined action of the **diaphragm** and the **intercostal muscles**.

The major muscle of respiration is the diaphragm. During inspiration the dome-shaped diaphragm contracts and flattens, increasing the volume of the thoracic cavity. It is responsible for 75 per cent of air movement into the lungs. The external intercostal muscles are also involved in respiration, and upon contraction they increase the depth of the thoracic cavity by pulling the ribs upwards and outwards. They are responsible for bringing approximately 25 per cent of the volume of air into the lungs. The combined contraction of the diaphragm and the external intercostals increase the thoracic cavity, which then decreases the pressure inside the thorax so that air from outside of the body enters the lungs.

Other accessory muscles which assist in inspiration include the sternomastoid, serratus anterior, pectoralis minor, pectoralis major and the scalene muscles in the neck.

During normal respiration, the process of expiration is passive and is brought about by the relaxation of the diaphragm and the external intercostal muscles, along with the elastic recoil of the lungs. This increases the internal pressure inside the thorax so that air is pushed out of the lungs.

KEY NOTE

Breathing is a relatively passive process. However, when more air needs to be exhaled (such as when coughing, or playing a wind instrument), the process of expiration becomes active. This is assisted by muscles such as the internal intercostals which help to depress the ribs. Abdominal muscles such as the external and internal obliques, rectus abdominus and the transversus abdominus help to compress the abdomen and force the diaphragm upwards, thus assisting expiration and squeezing more air out of the lungs.

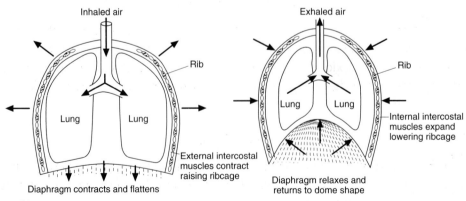

Figure 61
The mechanism of respiration

The Importance of Correct Breathing

Exercise increases the rate and depth of breathing due to the muscle cells requiring more oxygen. The breathing rate can more than double during vigorous exercise. Correct breathing is very important as it ensures that all the body's cells receive an adequate amount of oxygen and dispose of enough carbon dioxide to enable them to function efficiently.

It is important to note that breathing affects both our physiological and psychological state; deep breathing exercises can help to increase the vital capacity and function of the lungs.

Disorders of the Respiratory System

Asthma

A condition in which there are attacks of shortness of breath and difficulty in breathing due to spasm or swelling of the bronchial tubes. This is caused by hypersensitivity to allergens such as pollens of various plants, grass, flowers, pet hair, dust mites, and various proteins in foodstuffs such as shellfish, eggs and milk.

Asthma may be exacerbated by exercise, anxiety, stress or smoking. It runs in families and can be associated with hayfever and eczema.

Bronchitis

A chronic or acute inflammation of the bronchial tubes. Chronic bronchitis is common in smokers and may lead to emphysema, which is caused by damage to lung structure. Acute bronchitis can result from a recent cold or flu.

Emphysema

This is a chronic obstructive pulmonary disease in which the alveoli of the lungs become enlarged and damaged, reducing the surface area for the exchange of oxygen and carbon dioxide. Severe emphysema causes breathlessness which is made worse by infection. It is commonly associated with chronic bronchitis, smoking and advancing age.

Hay Fever

An allergic reaction involving the mucous passages of the upper respiratory tract and the conjunctiva of the eyes, caused by pollen or other allergens. Causes nose blockages, sneezing and watery eyes.

Pleurisy

Inflammation of the pleura of the lung. It presents as an intense stabbing pain over the chest on breathing deeply. There is difficulty in breathing, respiration is shallow and rapid, and fever is present. Pleurisy may develop as a complication of pneumonia, tuberculosis or trauma to the chest.

Pneumonia

Inflammation of the lung caused by bacteria, in which the alveoli become filled with inflammatory cells and the lung becomes solid. Symptoms include fever, malaise, headache together with a cough and chest pain.

Sinusitis

A condition involving inflammation of the paranasal sinuses. It is usually caused by a viral or bacterial infection, or may be associated with a common cold or allergy. The congestion of the nose results in a blockage in the opening of the sinus into the nasal cavity and a build-up of pressure in the sinus.

The condition presents with nasal congestion followed by a mucous discharge from the nose. The pain is located to specific areas depending on the sinuses affected. If the frontal sinuses are affected, a major symptom is a headache over one or both eyes.

If the maxillary sinuses are affected, one or both cheeks will hurt and it may feel as if there is a toothache in the upper jaw.

SELF ASSESSMENT QUESTIONS – THE RESPIRATORY SYSTEM

1. List the main respiratory structures.

....

....

....

2. State two functions of the respiratory system.

....

....

3. List the functions of the nose.

4. What is the pharynx and what is its function?

5. Describe the structure and function of the larynx.

6. State the functional significance of

a) the trachea

b) the bronchi.

7. Describe the anatomical position and structure of the lungs.

8. What is the significance of the alveoli in the lungs?

9. Give a brief description of the process of interchange of gases in the lungs.

10. Describe the mechanism of respiration that causes air to be drawn in and out of the lungs.

11. What is the importance of correct breathing?

12. State two factors that may affect the breathing rate.

CHAPTER 12

The Olfactory System

Olfaction is a special sense, which is capable of detecting different smells and evoking emotional responses due to its close link with the endocrine system (see Chapter 14). The process of olfaction is assisted by the nervous system, as smells received by the nose are transmitted by nerve impulses to be perceived by the brain.

A competent therapist needs to be able to have knowledge of the olfactory system in order to understand the process of olfaction.

By the end of this chapter you will be able to relate the following knowledge to your practical work carried out in the salon:

▶ the internal structures of the nasal cavity, along with the functional significance
▶ the theory of olfaction.

The Structure of the Olfactory System

The special features of the olfactory system are as follows:

▶ **Nose:** this is the organ of olfaction or smell.
▶ **Mucous membrane:** lines the nose, moistens the air passing over it and helps to dissolve the odorous gas passing through the nasal cavity. The mucous membrane has a very rich blood supply, and warmth from the blood flowing through the tiny capillaries in the nose raises the temperature of the air as it passes through the nose.

▶ **Cilia:** the tiny hairs inside the nose. They are highly sensitive and are extensions of nerve fibres connecting with the olfactory cells. The tips of the cilia are covered with mucous and they are able to detect tiny chemical odorous particles which enter the nose.

▶ **Olfactory cells:** these lie embedded in the mucous in the upper part of the nasal cavity. These nerve cells are sensory and are specially adapted for sensing smell. Each olfactory cell has a long nerve fibre called an **axon**, leading out of the main body of the cell, which picks up information received and passes it on to the brain.

▶ **Olfactory bulb:** the area of the brain, situated in the cerebral cortex, which perceives smell.

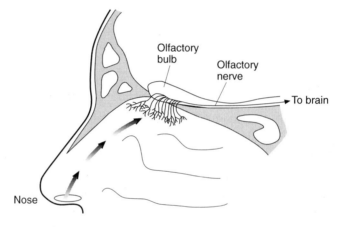

Figure 62
Internal structure of the nose

The Theory of Olfaction

▶ Particles of solid or liquid (essential oil) evaporate on contact with the air.

▶ Mucous membrane, lining the nose, dissolves odorous particles by warming them and mixing them with water vapour as they pass through the nose.

▶ Special cilia pass on to olfactory cells whatever information they have picked up about the evaporated gas passing through the nose.

▶ Nerve fibres of olfactory cells pass through a bony plate at the top of the nose and connect directly with the area of the brain known as the olfactory bulb.

▶ Smell is perceived by olfactory cells which connect directly with the olfactory bulb in the brain.

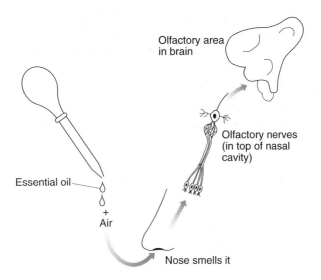

Figure 63
The process of olfaction

KEY NOTE

In most nerves in the body, the transmission of a nerve impulse is achieved through the spinal cord and then on to the brain. However, in the case of the olfactory cells, the nerve fibres connect *directly* with the olfactory bulb of the brain and therefore have a powerful and immediate effect on the emotions.

This can be explained by the fact that the area of the brain associated with smell is very closely connected with that part of the brain known as the limbic system, which is concerned with emotions, memory and sex drive. The olfactory bulb also connects closely with the hypothalamus, the nerve centre which governs the endocrine system.

Essential oils enter the nose in the form of gases, as they evaporate when in contact with air and are volatile in nature. We inhale them in this evaporated form.

1. List the individual parts which constitute the olfactory system

2. What is the functional significance of the olfactory cells and where are they located?

3. Briefly describe the process of olfaction

CHAPTER 13

The Nervous System

The anatomical structures of the nervous system include the brain, spinal cord and nerves which together form the main communication system for the body. The nervous system is the body's control centre or 'Head Office', and is therefore responsible for receiving and interpreting information from inside and outside the body.

The nervous system receives, interprets and integrates all stimuli to effect a response. It is also responsible for all mental processes and emotional responses, and works intimately with the endocrine system to help regulate body processes.

A competent therapist needs to be able to understand the principles of operation of the nervous system in order to carry out treatments safely and effectively.

By the end of this chapter you will be able to relate the following information to your practical work carried out in the salon:

▶ the characteristics of nervous tissue
▶ the different types of neurones
▶ the transmission of nerve impulses
▶ the organisation of the nervous system
▶ an outline of the principal parts of the nervous system
▶ disorders of the nervous system.

The nervous system has two main parts which both possess unique structural and functional characteristics.

1 The **Central Nervous System** (the main control system) that consists of the brain and the spinal cord
2 The **Peripheral Nervous System** consisting of 31 pairs of spinal nerves, 12 pairs of cranial nerves and the autonomic nervous system.

Nervous Tissue

Nervous tissue is composed of nerve cells called **neurones**, and also of **neuroglia** (the special connective tissue of the central nervous system that is designed to support, nourish and protect the neurones). The neuroglia are smaller than the neurones and are unable to transmit impulses.

Neurone

The functional unit of the nervous system is a neurone which is a specialised nerve cell, designed to receive stimuli and conduct impulses.

The nervous system contains billions of interconnecting neurones which are the basic impulse-conducting cells of the nervous system. Neurones have two major properties:

1 **excitability:** the ability to respond to a stimulus and convert it to a nerve impulse.
2 **conductability:** the ability to transmit the impulses to other neurones, muscles and glands.

Parts of a Neurone

Although neurones vary in their shape and size, they all have three basic parts: a cell body and two or more extensions, namely axons and dendrites.

▶ The cell body contains the nucleus and other standard organelles of the cells.
▶ **Dendrites** are highly branched extensions of the nerve cell. These neural extensions receive and transmit stimuli *towards* the cell body.
▶ **Axons** are typically cylindrical extensions of the cell. Their function is to transmit impulses *away* from the cell body.

There are three types of neurones:

1 **Sensory** or **afferent** neurones receive stimuli from sensory organs and receptors, and transmit the impulse to the spinal cord and brain. Sensations transmitted by the sensory neurones include heat, cold, pain, taste, smell, sight and hearing.

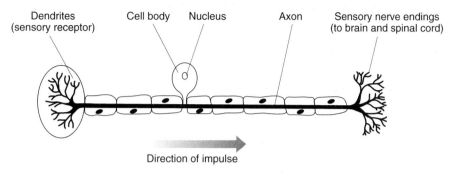

Figure 64
Sensory nerve

2 **Motor** or **efferent** neurones conduct impulses away from the brain and the spinal cord to muscles and glands, in order to stimulate them into carrying out their activities.

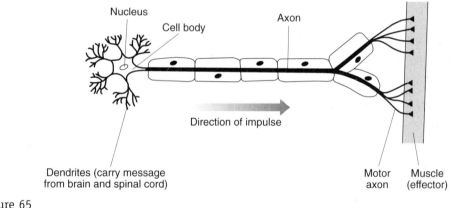

Figure 65
Motor nerve

3 **Association (mixed) neurones:** these link sensory and motor neurones, helping to form the complex pathways that enable the brain to interpret incoming sensory messages, decide on what should be done, and send out instructions in response along motor pathways, to keep the body functioning properly.

The Transmission of Nerve Impulses

Nerve impulses are the signals of the nervous system that travel along the neurone from dendrite to axon. The function of a neurone is to transmit impulses from their origin to destination. The nerve fibres of a neurone are not actually joined together and therefore there is no anatomical continuity between one neurone and another.

The junction where nerve imulses are transmitted from one neurone to another is called a **synapse**; this is the junction between two neurones or between a neurone and a muscle or gland, where they connect to transmit information.

Impulses are relayed from one neurone to another by a chemical transmitter substance, which is released by the neurone to carry impulses across the synapse to stimulate the next neurone. Synapses cause nerve impulses to pass in one direction only and are important in coordinating the actions of neurones. A special kind of synapse occurs at the junction between a nerve and a muscle and is known as a **motor point**, which is the point where the nerve supply enters the muscle.

The conduction of motor impulse in the contraction of voluntary muscle:

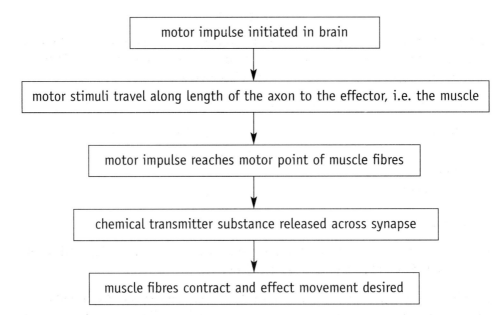

motor impulse initiated in brain

motor stimuli travel along length of the axon to the effector, i.e. the muscle

motor impulse reaches motor point of muscle fibres

chemical transmitter substance released across synapse

muscle fibres contract and effect movement desired

KEY NOTE

An important neurotransmitter is **acetylcholine** which is vital to muscle contraction.

The Central Nervous System

The central nervous system, consisting of the brain and spinal cord, is covered by a special type of connective tissue called the **meninges.** The meninges has three layers:

1   the **dura mater**, which is the outer protective fibrous connective tissue sheath
    covering the brain and spinal cord.
2   the **pia mater**, the innermost layer, which is attached to the surface of organs
    and is richly supplied with blood vessels to nourish the underlying tissues.
3   the **arachnoid mater** which provides a space for the blood vessels and
    circulation of cerebrospinal fluid.

Cerebrospinal fluid is a clear fluid derived from the blood and secreted into the
inner cavities of the brain. It carries some nutrients to the nerve tissue and carries
waste away, but its main function is to protect the central nervous system by acting
as a shock absorber for the delicate nervous tissue.

The Brain

The brain is an extremely complex mass of nervous tissue lying within the skull. It
is the main communication centre of the nervous system and its function is to
coordinate the nerve stimuli received and effect the correct responses.

The main parts of the brain include the:

▶ cerebrum
▶ thalamus
▶ hypothalamus
▶ pituitary gland

▶ pineal gland
▶ cerebellum.

The Cerebrum

This is the largest portion of the brain and makes up the front and top part of the
brain. It is divided into two large cerebral hemispheres.

The outer layer of the cerebrum is called the **cerebral cortex** and is the region
where the main functions of the cerebrum are carried out. The cortex is concerned
with all forms of conscious activity: sensations such as vision, touch, hearing, taste
and smell; control of voluntary movements; reasoning; emotion; and memory. The
cortex of each cerebral hemisphere has a number of functional areas:

▶ **sensory areas** – these receive impulses from sensory organs all over the body;
  there are separate sensory areas for vision, hearing, touch, taste and smell.
▶ **motor areas** – these areas have motor connections through motor nerve fibres
  with voluntary muscles all over the body.
▶ **association areas** – in these areas association takes place between information
  from the sensory areas and remembered information from past experiences.
  Conscious thought then takes place, and decisions are made which often result in
  conscious motor activity controlled by motor areas.

KEY NOTE

The brain requires a continuous supply of glucose and oxygen as it is unable to store glycogen, unlike the liver and muscles.

The Thalamus

This is a relay and interpretation centre for all sensory impulses, except olfaction.

The Hypothalamus

This small structure governs many important homeostatic functions. It regulates the autonomic nervous and endocrine systems by governing the pituitary gland. It controls hunger, thirst, temperature regulation, anger, aggression, hormones, sexual behaviour, sleep patterns and consciousness.

The Pituitary Gland

This is a pea-shaped body attached beneath the hypothalamus in a bony cavity at the base of the skull. It is known as the master endocrine gland, as its hormones control and stimulate other glands to produce their hormones.

The Pineal Gland

This is a pea-sized mass of nerve tissue attached by a stalk in the central part of the brain. It is located deep between the cerebral hemispheres, where it is attached to the upper portion of the thalamus. The pineal gland secretes a hormone called melatonin, which is synthesised from serotonin.

The pineal gland is involved in the regulation of circadian rhythms – patterns of repeated activity that are associated with the environmental cycles of day and night such as sleep / wake rhythms. The pineal gland is also thought to influence the mood.

The Cerebellum

The cerebellum is a cauliflower-shaped structure located at the posterior of the cranium, below the cerebrum. The cerebellum is concerned with muscle tone, the co-ordination of skeletal muscles and balance.

The Brain Stem contains three main structures:

1 the **mid-brain** – this contains the main nerve pathways connecting the cerebrum and the lower nervous system. It also contains certain visual and auditory reflexes that coordinate head and eye movements with things seen and heard.
2 the **pons** – this is below the mid-brain and relays messages from the cerebral cortex to the spinal cord.
3 the **medulla oblongata** – this is often considered the most vital part of the brain. It is an enlarged continuation of the spinal cord and connects the brain with the spinal cord. Control centres within the medulla oblongata include those for the heart, lungs and intestines.

TASK 1

Label the principal parts of the brain in Figure 66:

cerebrum   pituitary gland   cerebellum   brain stem (mid-brain, pons,

pineal gland   thalamus   hypothalamus   medulla oblongata)

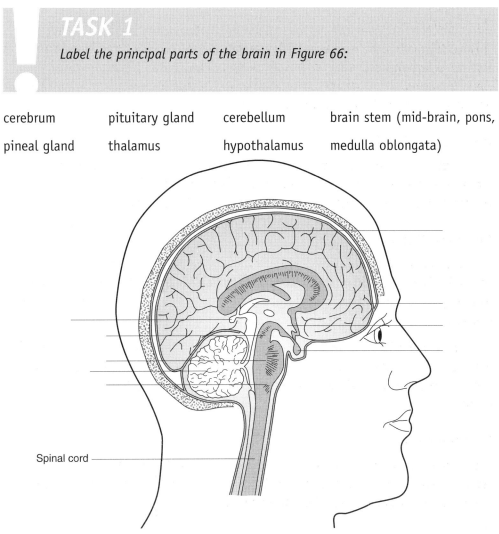

Spinal cord

Figure 66
Principal parts of the brain

The Blood-Brain Barrier

The blood-brain barrier is a selective semipermeable wall of blood capillaries with a thick basement membrane. It prevents, or slows down, the passage of some drugs and other chemical compounds, and keeps disease-causing organisms such as viruses from travelling into the central nervous system via the bloodstream.

The Spinal Cord

This is an extension of the brain stem, which extends from an opening at the base of the skull down to the second lumbar vertebra. Its function is to relay impulses to and from the brain; **sensory** tracts conduct impulses *to* the brain and **motor tracts** conduct impulses *from* the brain.

The spinal cord provides the nervous tissue link between the brain and other organs of the body, and is the centre for reflex actions which provide a fast response to external or internal stimuli.

KEY NOTE

A reflex action is a rapid and automatic response to a stimulus, without any conscious thought of the brain. A typical example of a reflex action is the knee-jerk response, which involves sensory and motor nerves being coordinated through the spinal cord.

The Peripheral Nervous System

The peripheral nervous system consists of cable-like nerves that link the central nervous system to the rest of the body. It contains:

▶ 31 pairs of spinal nerves
▶ 12 pairs of cranial nerves
▶ the autonomic nervous system.

Thirty-one Pairs of Spinal Nerves

These nerves pass out of the spinal cord, and each has two thin branches which link it with the autonomic nervous system. Spinal nerves receive sensory impulses from the body and transmit motor signals to specific regions of the body, thereby providing two-way communication between the central nervous system and the body.

Each of the spinal nerves are numbered and named, according to the level of the spinal column from which they emerge. There are:

- eight cervical
- twelve thoracic
- five lumbar
- five sacral
- one coccygeal spinal nerve.

Each spinal nerve is divided into several branches, forming a network of nerves or plexuses which supply different parts of the body:

- The **cervical plexuses** of the neck supply the skin and muscles of the head, neck, and upper region of the shoulders.
- The **brachial plexuses** are at the top of the shoulder and they supply the skin and muscles of the arm, shoulder and upper chest.
- The **lumbar plexuses** are located between the waist and the hip. They supply the front and sides of the abdominal wall and part of the thigh.
- The **sacral plexuses** at the base of the abdomen supply the skin and muscles and organs of the pelvis.
- The **coccygeal plexus** supplies the skin in the area of the coccyx, and the muscles of the pelvic floor.

Twelve Pairs of Cranial Nerves

These nerves connect directly to the brain. Between them, they provide a nerve supply to sensory organs, muscles and skin of the head and neck. Some of the nerves are mixed, containing both motor and sensory nerves, while others are either sensory or motor.

1 **Olfactory** – a sensory nerve of olfaction (smell).

2 **Optic** – a sensory nerve of vision.

3 **Oculomotor** – a mixed nerve that innervates both internal and external muscles of the eye and a muscle of the upper eyelid.

4 **Trochlear** – the smallest of the cranial nerves. It is a motor nerve that innervates the superior oblique muscle of the eyeball (which helps you look upwards).

5 **Trigemenal** – a mixed nerve (containing motor and sensory nerves) that conducts impulses to and from several areas in the face and neck. It also controls the muscles of mastication (the masseter, the temporalis and the pterygoids).

It has three main branches:

> The **opthalmic branch** carries sensations from the eye, nasal cavity, skin of forehead, upper eyelid, eyebrow and part of the nose.
> The **maxillary branch** carries sensations from the lower eyelid, upper lip, gums, teeth, cheek, nose, palate and part of the pharynx.
> The **mandibular branch** carries sensations from the lower gums, teeth, lips, palate and part of the tongue.

6 **Abducens** – a mixed nerve that innervates only the lateral rectus muscle of the eye, which helps you look to the side.

7 **Facial** – a mixed nerve that conducts impulses to and from several areas in the face and neck. The sensory branches are associated with the taste receptors on the tongue, and the motor fibres transmit impulses to the muscles of facial expression.

8 **Vestibulocochlear** – a sensory nerve that transmits impulses generated by auditory stimuli, and stimuli related to equilibrium, balance and movement.

9 **Glossopharyngeal** – a mixed nerve that innervates structures in the mouth and throat. It supplies motor fibres to part of the pharynx and to the parotid salivary glands, and sensory fibres to the posterior third of the tongue and the soft palate.

10 **Vagus** – unlike the other cranial nerves, it has branches to numerous organs in the thorax and abdomen as well as the neck. It supplies motor nerve fibres to the muscles of swallowing and to the heart and organs of the chest cavity. Sensory fibres carry impulses from the organs of the abdominal cavity and the sensation of taste from the mouth.

11 **Accessory** – functions primarily as a motor nerve, innervating muscles in the neck and upper back (such as the trapezius and the sternomastoid), as well as muscles of the palate, pharynx and larynx.

12 **Hypoglossal** – a motor nerve that innervates the muscles of the tongue.

The Autonomic Nervous System

This is the part of the nervous system that controls the automatic body activities of smooth and cardiac muscle and the activities of glands. It is divided into the **sympathetic** and **parasympathetic** divisions, which possess complementary responses.

The Sympathetic System

The activity of the sympathetic system is to prepare the body for expending energy and dealing with emergency situations. Its effects include:

- increased heart rate
- increased respiration rate
- dilation of skeletal blood vessels
- stimulation of the adrenal and sweat glands
- increased conversion of glycogen to glucose by liver
- pupil dilation
- decreased secretion of saliva
- relaxation of the bladder wall and the closing of the sphincter muscles
- decreased gastrointestinal activity.

KEY NOTE

The sympathetic stimulation of the autonomic nervous system is increased by the release of the hormone **adrenaline** from the adrenal medulla. This is an example of the nervous and endocrine system working synergistically.

The Parasympathetic System

This balances the action of the sympathetic division by working to conserve energy and create the conditions needed for rest and sleep. It slows down the body processes except digestion and the functions of the genito-urinary system. In general, the actions of the parasympathetic system oppose those of the sympathetic system, and the two systems work together to regulate the internal workings of the body.

Effects of the parasympathetic nervous system include:

- resting heart rate
- resting respiratory rate
- constriction of skeletal blood vessels
- increased gastro-intestinal activity (increased peristalsis and increased secretion of insulin and digestive juices)
- contraction of the bladder and the opening of the sphincter muscles
- pupil constriction
- stimulated salivation.

The sympathetic and parasympathetic nervous systems are finely balanced to ensure the optimum functioning of organs of the body.

TASK 2

Complete the table to show the differences in effect between the sympathetic and parasympathetic nervous systems.

Organ / Part of Body	Effects of Sympathetic Stimulation	Effects of Parasympathetic Stimulation
Heart	increases heart beat	
Bronchi		slows down
Skeletal blood vessels	dilation	
Pupils of the eyes		constriction
Sweat glands	increases	
Digestive system	decreases	
Bladder	relaxes bladder and closes sphincter muscles	

Disorders of the Nervous System
Anxiety

This can be defined as fear of the unknown, but as an illness it can vary from a mild form to panic attacks and severe phobias that can be disabling socially, psychologically and at times physically.

It presents with a feeling of dread that something serious is likely to happen, and is associated with palpitations, rapid breathing, sweaty hands, tremor (shakiness), dry mouth, general indigestion, feeling of butterflies in the stomach, occasional diarrhoea and generalised aches and pains in the muscles. It can present with similar features of mild to moderate depression of the agitated type. The causes of anxiety can be related to personality with some genetic and behavioural predisposition, or a traumatic experience or physical illness, ie hyperthyroidism.

Bells Palsy

A disorder of the seventh cranial nerve (facial nerve) that results in paralysis on one side of the face. The disorder usually comes on suddenly and is commonly caused by inflammation around the facial nerve as it travels from the brain to the exterior.

It may be caused by pressure on the nerve due to tumours, injury to the nerve, infection of the meninges or inner ear, or dental surgery. Diabetes, pregnancy and hypertension are other causes.

The condition may present with a drooping of the mouth on the affected side due to flaccid paralysis of the facial muscles, and there may be difficulty in puckering the lips due to paralysis of the orbicularis oris muscle.

▶ Taste may be diminished or lost if the nerve has been affected proximal to the branch, which carries taste sensations.
▶ It may be difficult to close the eye tightly and crease the forehead.
▶ The buccinator muscle is affected, which prevents the client from puffing the cheeks and is the cause of food getting caught between the teeth and cheeks.
▶ There is also excessive tearing from the affected eye.
▶ Pain may be present near the angle of the jaw and behind the ear.

Between 80 and 90 per cent of individuals recover spontaneously and completely in around one to eight weeks. Corticosteriods may be used to reduce the inflammation of the nerve.

Carpal Tunnel Syndrome

This syndrome is characterised by pain and numbness in the thumb or hand, resulting from pressure on the median nerve of the wrist. Pain and 'pins and needles' sensation may radiate to the elbow. It is known to cause severe pain at night and can cause muscle wasting of the hand.

There is a higher risk of this condition in occupations requiring repetitive strains of the wrist, such as massage therapists and secretaries.

Cerebral Palsy

A condition caused by damage to the central nervous system, of the baby during pregnancy, delivery or soon after birth. The damage could be due to bleeding, lack of oxygen or other injuries to the brain.

The signs and symptoms of this condition depend on the area of the brain affected.

▶ Speech is impaired in most individuals, and there may be difficulty in swallowing.
▶ There may or may not be mental retardation.
▶ Muscles may increase in tone to become spastic, making coordinated movements difficult. The muscles are hyperexcitable and even small movements, touch, stretch of muscle or emotional stress can increase the spasticity.
▶ The posture is abnormal due to muscle spasticity, and the gait is also affected.

Some may have abnormal involuntary movements of the limbs that may be exaggerated on voluntarily performing a task. Weakness of muscles may also be associated with the condition, along with seizures.

▶ There may be problems with hearing and vision.

Depression

This combines symptoms of lowered mood, loss of appetite, poor sleep, lack of concentration and interest, lack of sense of enjoyment, occasional constipation and loss of libido. There are occasions when there is suicidal thinking, death wish or active suicide attempts.

Depression can be the result of chemical imbalance, usually related to serotonin and noradrenalin. The cause of depression could be endogenous where there is no cause for depression, but is thought to be linked to genetic predisposition, the result of physical illness, or loss of a close relative, object, limb or a relationship. A depressed person looks miserable, hunchbacked, downcast and will usually avoid eye contact.

The severity, as suggested above, can be variable, but may become severe enough to become psychotic manifested by hallucinations, delusions, paranoia or thought disorders.

Epilepsy

A neurological disorder which makes the individual susceptible to recurrent and temporary seizures. Epilepsy is a complex condition, and classifications of types of epilepsy are not definitive.

Generalised

This may take the form of major or tonic-clonic seizures (formerly known as *grand mal*) in which, at the onset, the patient falls to the ground unconscious with their muscles in a state of spasm (tonic phase). This is then replaced by convulsive movements (the clonic phase), when the tongue may be bitten and urinary incontinence may occur. Movements gradually cease and the patient may rouse in a state of confusion, complaining of a headache, or may fall asleep.

Partial

This may be idiopathic or a symptom of structural damage to the brain. In one type of partial idiopathic epilepsy, often affecting children, seizures may take the form of absences (formerly known as *petit mal*), in which there are brief spells of unconsciousness lasting for a few seconds. The eyes stare blankly and there may be fluttering movements of the lids and momentary twitching of the fingers and mouth.

This form of epilepsy seldom appears before the age of three or after adolescence. It often subsides spontaneously in adult life, but may be followed by the onset of generalised or partial epilepsy.

Focal

This is partial epilepsy due to brain damage (either local or due to a stroke). The nature of the seizure depends on the location of the damage in the brain. In a Jacksonian motor seizure, the convulsive movements may spread from the thumb to the hand, arm and face.

Psychomotor

This type of epilepsy is caused by dysfunction of the cortex of the temporal lobe of the brain. Symptoms may include hallucinations of smell, taste, sight and hearing. Throughout an attack the patient is in a state of clouded awareness, and afterwards may have no recollection of the event.

Headache

Pain which affects the head, excluding facial pain. It can result from diseases affecting ear, nose and throat, for example sinusitis, as well as eye problems which could be corrected by glasses. Types of headaches include:

- **Simple headache** – this may occur at times of stress, during periods, the day after heavy alcohol consumption, part of cold and flu symptoms. These are transient and would normally settle spontaneously or require simple analgesia.
- **Chronic headaches** – daily headache and tension headache. The pain can be severe and disabling and can affect the whole head, behind the eyes or may be just a frontal headache. The client can describe the pain as like a band around the head.
- **Headache due to radiation from cervical spines (Cervicalgia)** – this is normally in the back and sides of the head, and can present with neck pain.
- **Migraine headache** – see Migraine overleaf.
- **Headaches due to Intracranial (inside brain) diseases** – headaches caused by diseases such as brain tumour, can present with nausea and vomiting, and may cause other neurological signs and symptoms.

Herpes Zoster (shingles)

Painful infection along the sensory nerves by the virus that causes chicken pox. Lesions resemble herpes simplex with erythema and blisters along the lines of the nerves. Areas affected are mostly on the back or upper chest wall.

This condition is very painful due to acute inflammation of one or more of the peripheral nerves. Severe pain may persist at the site of shingles for months or even years after the apparent healing of the skin.

Meningitis

Inflammation of the meninges due to infection by viruses or bacteria. Meningitis presents with an intense headache, fever, loss of appetite, intolerance to light and sound, and rigidity of muscles, especially those in the neck. In severe cases there may be convulsions, vomiting and delirium leading to death.

▶ In **meningocccal** meningitis (involving a characteristic haemorrhagic rash anywhere on the body) the symptoms appear suddenly, and the bacteria can cause widespread meningococcal infection culminating in meningococcal septicaemia. Unless treated rapidly, death can occur within a week.
▶ **Bacterial** meningitis is treated with large doses of antibiotics.
▶ **Viral** meningitis does not respond to drugs but normally has a relatively benign prognosis.

Migraine

Specific form of headache, usually unilateral (one side of the head), associated with nausea or vomiting, visual disturbances such as scintillating light waves or zigzag fashion. Client may experience a visual aura before an attack actually happens. This is usually called a **classical migraine**. There are other types of migraine:

▶ **Ophthalmoplegic migraine** causes painful, red and watery eyes
▶ **Neuropathic migraine** causes one-sided paralysis, and weakness of the face and body.
▶ **Abdominal migraine** can affect children, with recurring attacks of abdominal pain, sometimes accompanied by nausea / vomiting.

Migraine can be treated with simple analgesia or more specialised anti-migraine medication.

Motor Neurone Disease

A progressive degenerative disease of the motor neurones of the nervous system. It tends to occur in middle age, and causes muscle weakness and wasting.

Multiple Sclerosis

Disease of the central nervous system, in which the myelin (fatty) sheath covering the nerve fibres is destroyed and various functions become impaired, including movement and sensations. Multiple sclerosis is characterised by relapses and remissions.

It can present with blindness or reduced vision and can lead to severe disability within a short period. It can also cause incontinence, loss of balance, tremor and speech problems. Depression and mania can happen.

Parkinson's Disease

Damage to grey matter of brain known as **basal ganglia**. Causes involuntary tremors of limbs, with stiffness, rigidity, shuffling gait. Face lacks expression and movements are slow. Clients may suffer from depression, confusion and anxiety.

Sciatica

Lower back pain which can affect the buttock and thigh. On occasions it radiates to the leg and foot. In severe cases it can cause numbness and weakness of the lower limb. It can result from prolapse of the discs between the spinal vertebrae, tumour or blood clot (thrombosis). Diabetes or heavy alcohol intake can also produce symptoms of sciatica. This condition tends to recur and may require strong analgesia or surgery in severe cases.

SELF ASSESSMENT QUESTIONS – THE NERVOUS SYSTEM

1. How is the nervous system organised? List the classifications and their relevant parts.

2. State the characteristics of nervous tissue.

3. State the three basic parts of a neurone.

4. What is the difference between a sensory and a motor neurone?

5. What is meant by the term 'motor point' in relation to the transmission of a nerve impulse?

6. Briefly describe how a motor impulse is involved in the contraction of voluntary muscle.

7. Identify the following parts of the brain from the description given:

a) the part of the brain that is concerned with all forms of conscious activity: vision, touch, hearing, taste, smell, control of voluntary movements, reasoning learning and memory

b) the part that is concerned with the coordination of skeletal muscle and balance

c) the part that acts as a relay centre for all sensory impulses (except olfaction)

d) the part that relays messages from the cerebral cortex to the spinal cord

e) the part that contains vital control centres for the heart, lungs and intestines

8. What is the function of the spinal cord?

9. State the structures forming:

a) the autonomic nervous system

b) the peripheral nervous system

CHAPTER 14

The Endocrine System

The endocrine system comprises a series of internal secretions called *hormones* which help to regulate body processes by providing a constant internal environment. Hormones are chemical messengers, and act as catalysts in that they affect the physiological activities of other cells in the body. The endocrine system works closely with the nervous system; nerves enable the body to respond rapidly to stimuli, whereas the endocrine system causes slower and longer-lasting effects.

A competent therapist needs to understand the action of hormones and their significance in the healthy functioning of the body, as over-secretion and under-secretion of hormones may result in certain disorders and disease.

By the end of this chapter, you will be able to relate the following knowledge to your pratical work carried out in the salon:

▶ what a hormone is
▶ the location of the main endocrine glands of the body
▶ the principal secretions of the main endocrine glands
▶ the effects of hormones on the body
▶ disorders of the endocrine system.

The functions of the endocrine system are:

1 producing and secreting hormones which regulate body activities such as growth, development and metabolism

2 maintaining the body during times of stress

3 contributing to the reproductive process.

What is a Hormone?

A hormone is a chemical messenger or regulator, secreted by an endocrine gland, which reaches its destination by the bloodstream, and has the power of influencing the activity of other organs. Some hormones have a slow action over a period of years, for example the growth hormone from the anterior pituitary, while others have a quick action such as adrenaline from the adrenal medulla. Hormones therefore regulate and coordinate various functions in the body.

The endocrine glands are ductless glands, as the hormones they secrete pass directly into the bloodstream to influence the activity of another organ or gland. The main endocrine glands are as follows:

▶ the pituitary gland
▶ the thyroid gland
▶ the parathyroid glands
▶ the adrenal glands
▶ the islets of langerhans
▶ ovaries in the female
▶ testes in the male.

The Pituitary Gland

This is a lobed structure attached by a stalk to the hypothalamus of the brain. It is often referred to as the 'master gland' since it produces several hormones or *releasing factors* which influence the secretion of hormones by other endocrine organs. The pituitary gland consists of two main parts, an anterior and a posterior lobe.

The Anterior Lobe

The principal hormones secreted by the anterior lobe of the pituitary are as follows:

▶ **growth hormone**, which controls the growth of long bones and muscles
▶ **thyroid stimulating hormone (TSH)**, which controls the growth and activity of the thyroid gland
▶ **adrenocorticothrophic hormone (ACTH)**, which stimulates and controls the growth and hormonal output of the adrenal cortex
▶ **gonadotrophic hormones** control the development and growth of the ovaries and testes. The gonads or sex hormones include:

1 **follicle stimulating hormone**, which in women stimulates the development of the graafian follicle in the ovary which secretes the hormone oestrogen. In men it stimulates the testes to produce sperm.

2 **luteinizing hormone**, which in women helps to prepare the uterus for the fertilised ovum. In men, it acts on the testes to produce testosterone.

▶ **prolactin** stimulates the secretion of milk from the breasts following birth.

KEY NOTE

Endocrine glands in the body have a feedback mechanism which is coordinated by the pituitary gland. The pituitary gland is influenced by the hypothalamus, and will increase its output of releasing factors if other glands start to fail, or will decrease its output if the level of the hormone in the bloodstream starts to rise.

The Posterior Lobe

The posterior lobe of the pituitary secretes two hormones, which are manufactured in the hypothalamus but are stored in the posterior lobe:

▶ **the anti-diuretic hormone (ADH)** which increases water re-absorption in the renal tubules of the kidneys

▶ **oxytocin** stimulates the uterus during labour and stimulates the breasts to produce milk

▶ **melanocyte stimulating hormone (MSH)** is secreted by the central lobe of the pituitary gland and stimulates the production of melanin in the basal cell layer of the skin.

Disorders of the Pituitary Gland

Hypersecretion

Hypersecretion of the growth hormone secreted by the anterior pituitary leads to gigantism in children, a disease marked by the rapid growth of the body to extremely large proportions (7–8 ft).

If the overproduction occurs in adulthood, then there is abnormal enlargement of the hands and feet and coarsening of the facial features, due to the continued growth of tissues. This condition is known as acromegaly.

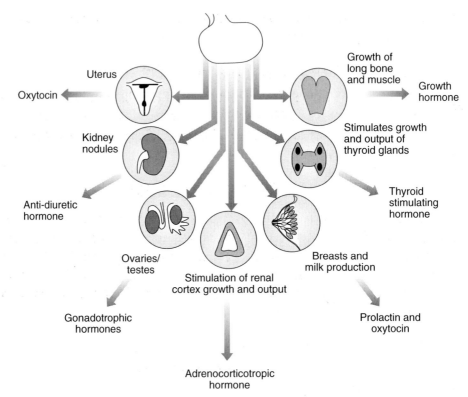

Figure 67
The pituitary and its master control

Hyposecretion

Hyopsecretion of the growth hormone during childhood leads to stunted growth, a condition known as dwarfism.

Hyposecretion of the anti-diuretic hormone by the posterior lobe of the pituitary leads to the disease Diabetes Insipidus. Symptoms include dehydration, an increased thirst and an increased output of urine.

The Pineal Gland

This is a pea-sized mass of nerve tissue attached by a stalk in the central part of the brain. It is located deep between the cerebral hemispheres, where it is attached to the upper portion of the thalamus.

The pineal gland functions as a gland and secretes a hormone called **melatonin**, which synthesises from serotonin. This gland is involved in the regulation of circadian rhythms, patterns of repeated activity that are associated with the environmental cycles of day and night such as sleep / wake rhythms. It is also thought to influence the mood.

Disorders of the Pineal Gland

Hyposecretion of the hormone melatonin is thought to be associated with the condition Seasonal Affective Disorder (SAD). Symptoms include depression (typically with the onset of winter), a general slowing down of mind and body, and excessive sleeping and overeating.

The Thyroid Gland

The thyroid gland is found in the neck, situated on either side of the trachea and is controlled by the anterior lobe of the pituitary. The principal secretions of the thyroid gland are **Triodothyronine (T3)** and **Thyroxine (T4)** which both regulate growth and development, as well as influence mental, physical and metabolic activities. The thyroid gland also secretes the hormone **calcitonin** which controls the level of calcium in the blood.

The functions of the thyroid gland are as follows:

- controls the metabolic rate by stimulating metabolism
- influences growth and cell division
- influences mental development
- is responsible for the maintenance of healthy skin and hair
- stores the mineral iodine, which it needs to manufacture thyroxin
- stimulates the involuntary nervous system and controls irritability.

The thyroid gland is controlled by a feedback mechanism. It will increase to meet the demand for more thyroid hormones at various times, such as during the menstrual cycle, at pregnancy, and puberty.

Disorders of the Thyroid Gland

Hypersecretion

Hypersecretion leads to a condition known as **thyrotoxicosis** or **Graves disease**. Thytroxicosis results in an increased metabolic rate, weight loss, sweating, restlessness, increased appetite, sensitivity to heat, raised temperature, frequent bowel action, anxiety and nervousness. When the thyroid gland produces and secretes an excessive amount of thyroxine, it may produce a goitre (an enlargement of the thyroid gland).

Hyposecretion

Hyposecretion leads to **cretinism** in children, which is a congential deficiency causing impaired mentality, small stature, coarsening of the skin and hair and deposition of fat on the body.

Hyposecretion in an adult leads to **myxoedema** which is characterised by the slowing down of physical and mental activity, with resultant lethargy, hair becomes brittle, skin becomes coarse and dry, and metabolism is slow.

The Parathyroid Glands

These are four small glands situated on the posterior of the thyroid gland. Their principal secretion is the hormone **parathormone** which helps to regulate calcium metabolism by controlling the amount of calcium in blood and bones.

Disorders of the Parathyroid Glands

Hypersecretion

Hypersecretion of parathormone, when there is an enlargement of the gland, causes re-absorption of calcium from bones, thus raising the blood calcium level. Effects may be renal stones, kidney failure, calcification of the soft tissues (softening of the bones) and tumours.

Hyposecretion

When there is a deficiency of calcium in the blood a condition known as tetany occurs, which is characterised by muscular twitchings particular of the hands and feet. These symptoms are quickly relieved by administering calcium.

The Adrenal Glands

These are two triangular shaped glands which lie on top of each kidney. They consist of two parts, an outer cortex and an inner medulla.

The Adrenal Cortex

The principal hormones secreted by the adrenal cortex are as follows:

1 **Gluco corticoids (cortisone and hydrocortisone):** these hormones influence the metabolism of protein and carbohydrates and utilise fats. They are important in maintaining the level of glucose in the blood, so that blood glucose levels are increased at times of stress.
2 **Mineral corticoids (aldosterone):** this hormone acts on the kidney tubules,

retaining salts in the body, excreting excess potassium and maintaining the water and electrolyte balance.

3 **Sex corticoids**: these include **testosterone, oestrogen** and **progesterone**: these hormones control the development of the secondary sex characteristics and the function of the reproductive organs.

Disorders of the Adrenal Glands

Hypersecretion of the Adrenal Cortex

Hypersecretion of the mineral corticoids can lead to kidney failure, high blood pressure and an excess of potassium in the blood causing an abnormal heartbeat. Hypersecretion of the glucorticoids can lead to a condition known as Cushings syndrome. This condition results from an excess amount of corticosteroid hormones in the body. Symptoms include weight gain, reddening of the face and neck, excess growth of facial and body hair, raised blood pressure, loss of mineral from bone and sometimes mental disturbances.

Hypersecretion of the sex corticoids may lead to hirsutism and amenorrhea in the female (due to hypersecretion of testosterone) and in the male may lead to muscle atrophy and the development of breasts (due to hypersecretion of oestrogen).

Hyposecretion of the Adrenal Cortex

Undersecretion of corticosteroid hormones is responsible for the condition known as Addison's disease. Symptoms include loss of appetite, weight loss, brown pigmentation around joints, low blood sugar, low blood pressure, tiredness, muscular weakness. This disease is treatable by replacement hormone therapy.

KEY NOTES

When the ovaries and testes mature, they produce the sex hormones themselves, therefore the production of sex corticoids in the adrenal cortex is important up to puberty.

The Adrenal Medulla

The principal hormones secreted by the adrenal medulla are **adrenaline** and **noradrenaline.** They are under the control of the sympathetic nervous system and are released at times of stress. The reponses of these hormones are fast due to the fact that they are governed by nervous control.

The effects of these stress hormones are similar, although adrenaline has a primary influence on the heart, causing an increase in heart rate, whereas noradrenaline has a greater effect on peripheral vasoconstriction, which raises blood pressure.

A summary of the effects of **adrenaline** are as follows:

- dilates the arteries, increasing blood circulation and the heart rate
- dilates the bronchial tubes, increasing oxygen intake and the rate and depth of breathing
- raises the metabolic rate
- constricts the blood vessels to the skin and intestines, diverting blood from these regions to your muscles and brain to effect action.

The effects of **noradrenaline** are similar to those of adrenaline and include:

- vasoconstriction of small blood vessels, leading to an increase in blood pressure
- increase in the rate and depth of breathing
- relaxation of the smooth muscle of the intestinal wall.

KEY NOTE

The effects described above are those felt when the body is under stress – a pounding heart, increased ventilation rate, a dry mouth and butterliles in the stomach.

Levels of stress hormones are broken down slowly so that effects on the sympathetic nervous system are long lasting. Over the long term, if levels of these hormones remain elevated, they perpetuate factors for stress-related disorders.

The Pancreas

The pancreas is known as a dual organ, as it has an endocrine and an exocrine function:

- The exocrine or external secretion is the secretion of pancreatic juice, to assist with digestion.
- The endocrine or internal secretion is the hormone insulin, secreted by the **islets of langerhans** cells in the pancreas. Insulin lowers the level of sugar in the blood by helping the body cells to take it up and use or store it as glycogen.

Disorders of the Islets of Langerhans in the Pancreas

Hypersecretion

This can lead to hypoglycaemia (low blood sugar level), causing muscular weakness and incoordination, mental confusion and sweating. If severe it may lead to a hypoglycaemic coma.

Hyposecretion

Hyposecretion can lead to a condition called Diabetes Mellitus. This condition is due to a deficiency or absence of insulin. The symptoms associated with diabetes include an increased thirst, increased output of urine, weight loss, thin skin with impaired healing capacity, increased tendency to develop minor skin infections, and decreased pain threshold when insulin levels are low.

There are two types of Diabetes Mellitus:

1 Insulin Dependent Diabetes (early onset)
   This occurs mainly in children and young adults, and the onset in usually sudden. The deficiency or absence of insulin is due to the destruction of the islet cells in the pancreas. The causes are unknown but there is a familial tendency, suggesting genetic involvement.

2 Non-insulin Dependent Diabetes (late onset)
   This type of diabetes occurs later in life and its causes are unknown. Insulin secretion may be below or above normal. Deficiency of glucose inside the body cells may occur where there is hyperglycaemia and a high insulin level. This may be due to changes in cell walls which block the insulin-assisted movement of glucose into cells. This type of diabetes can be controlled by diet alone, or diet and oral drugs.

The Sex Glands
The Testes

The testes are situated in the groin, in a sac called the scrotum. They have two functions:

▶ the secretion of the hormone **testosterone,** which controls the development of the secondary sex characteristics in the male at puberty (influenced by the luteinising hormone)
▶ the production of **sperm** (influenced by the follicle stimulating hormone from the anterior pituitary).

The ovaries

The ovaries are situated in the lower abdomen, below the kidneys. The two ovaries are the sex glands in the female, each is attached to the upper part of the uterus by broad ligaments. The ovaries have two distinct functions:

▶ the production of ova at ovulation
▶ production of the two hormones oestrogen and progesterone, which influence the secondary sex characteristics in the female and affect the process of reproduction.

Disorders of the Sex Hormones

Hypersecretion of the hormone testosterone in women can lead to **Virilism** (masculinisation) where there is an overproduction of androgens, **Hirsutism** (hair growth in the male sexual pattern) and **Amenorrhea** (the absence or stopping of periods).

Hypersecretion of oestrogen and progesterone in the male can lead to muscle atrophy and breast development (**Gynaecomastia).**

Hyopsecretion of the hormones oestrogen and progesterone in the female can lead to **Polycystic Ovary Syndrome**, which is characterised by cysts on the ovaries, cessation of periods, obesity, atrophy of the breasts, hirsutism and sterility.

Natural Glandular Changes
Puberty

The hormones oestrogen and progesterone become active at puberty and are responsible for the development of the secondary sex characteristics. The ovaries are stimulated by the gonadotrophic hormones from the anterior pituitary, known as the follicle stimulating hormone and the luteinising hormone. At puberty, oestrogen causes the breasts, the vulva and the vagina to gradually develop to mature proportions.

Oestrogen influences the amount of subcutaneous fat, determining the rounded contours of the female figure and is also concerned with the growth of the milk ducts in the breast. Oestrogen also influences the menstrual cycle and thickens the uterus lining in preparation for conception.

Pregnancy

Progesterone is known as the pregnancy hormone as it is concerned with the development of the placenta, which is a temporary endocrine gland during pregnancy. Progesterone helps with maintenance of the pregnancy and prepares the breasts for lactation.

Menopause

The menopause marks the end of a woman's reproductive life when oestrogen production begins to decline. The ovaries gradually become less responsive to the sex hormones, and hence ovulation and the menstrual cycles become irregular until they cease altogether.

KEY NOTE

Normally, hormones produced by the ovaries have an inhibitory or restraining effect on the anterior pituitary. However, in the case of the menopause, a lack of oestrogen results in a lack of proper control over this master gland, which then begins to pour out a flood of stimulating hormones.

This results in hyperstimulation by the pituitary hormones of the adrenal cortex, which in turn produces an excess of androgens or male hormones. It is for this reason that women of menopausal age, whose ovarian activity is declining, find themselves developing excess facial and body hair.

TASK 1 – MAIN ENDOCRINE GLANDS IN THE BODY

Label the main endocrine glands on Figure 68:

pituitary gland   thyroid gland   adrenal glands   pancreas

parathyroid glands   ovaries (female)   testes (male)

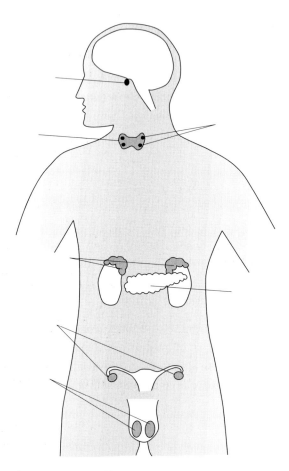

Figure 68
Main endocrine glands

TASK 2

Complete the following table to identify the hormonal secretions and the endocrine gland by which they are secreted.

Name of Hormone/s	Secreted by	Effects
		stimulates the uterus during labour
		controls the growth and activity of the thyroid gland
		control the growth and development of the ovaries and testes
		controls the levels of glucose in the blood
		prepares the uterus for pregnancy
		regulates calcium metabolism
		increases water re-absorption in the kidney tubules
		stimulates the secretion of milk from the breasts following birth

SELF ASSESSMENT QUESTIONS – THE ENDOCRINE SYSTEM

1. What is a hormone?

....

....

....

2. Name the main endocrine glands and their location in the body.

3. Why is the pituitary known as the master gland?

4. Identify the endocrine gland that secretes / releases the following hormone/s:

a) glucocorticoids

b) oxytocin

c) thyroid stimulating hormone

d) prolactin

e) parathormone

f) insulin

5. State the effects of the following hormones on the body:

a) follicle stimulating hormone

b) thyroxine and triodothyronine

c) adrenocorticotrophic hormone

d) adrenalin

....

....

e) testosterone

....

....

f) mineral corticoids

....

....

6. Briefly describe the effects the hormones oestrogen and progesterone have on the body during:

a) puberty

....

....

....

b) pregnancy

....

....

....

c) the menopause

....

....

....

7. State the endocrine disorders resulting from:

a) hypersecretion of thyroxine

....

....

....

b) hyposecretion of parathormone

....

....

....

c) hypersecretion of glucocorticoids

....

....

....

d) hyposecretion of oestrogen and progesterone

....

....

....

e) hyposecretion of melatonin

....

....

....

f) hyposecretion of thyroxine

....

....

g) hyposecretion of glucocorticoids

....

....

....

h) hypersecretion of the growth hormone

....

....

....

8. What is meant by the following terms and what may cause these to occur?

a) hirsutism

....

....

b) virilism

....

....

CHAPTER 15

The Reproductive System

Of all the body's systems, the reproductive system is significantly different between the two sexes. The male and female reproductive systems are specialised, in their dual function to produce the sex hormones responsible for the male and female characteristics, and for producing the cells required for reproduction.

These systems are unique in that they are not vital to the survival of an individual, but they are essential to the continuation of the human species.

By the end of this chapter you will be able to recall and understand the following in relation to your work as a therapist:

▶ an outline of the female reproductive system
▶ natural glandular changes that occur in the body – menstruation, pregnancy and menopause
▶ an outline of the male reproductive system.

The Female Reproductive System

The function of the female reproductive system is the production of sex hormones and ova (egg cells), which if fertilised, are supported and protected until birth. The female reproductive system consists of the following internal organs lying in the pelvic cavity:

▶ the ovaries
▶ the fallopian tubes
▶ the uterus
▶ the vagina.

The external genitalia is known collectively as the **vulva** and consists of:

▶ the labia major and minor, which are lip-like folds at the entrance of the vagina
▶ the clitoris, which is attached to the symphysis pubis by a suspensory ligament and contains erectile tissue
▶ the hymen, which is a thin layer of mucous membrane
▶ the greater vestibular glands which lie in the labia majora, one on each side near the vaginal opening. These glands secrete mucus which lubricates the vulva.

The breasts are accessory glands to the female reproductive system.

The Ovaries

These are the female sex glands and they lie on the lateral walls of the pelvis. They are almond-shaped organs which are held in place, one on each side of the uterus, by several ligaments. The largest of the ligaments is the broad ligament which holds the ovaries in close proximity to the fallopian tubes.

The ovary contains numerous small masses of cells called ovarian follicles, within which the ova (egg cells) develop. At the time of birth there are about two million immature ova in the ovaries. Many of the ova degenerate, and at the time of puberty there are only about 400,000 left.

The immature ova (or ocytes) lie dormant in the ovary until they are stimulated by a sudden surge in the hormone FSH (follicle stimulating hormone) at the time of puberty. Normally one egg (ovum) ripens and is released each month.

Functions

The ovaries have two distinct functions:

1 the production of ova
2 the secretion of the female hormones oestrogen and progesterone.

Oestrogen and progesterone regulate the changes in the uterus throughout the menstrual cycle and pregnancy. Oestrogen is responsible for the development of the female sexual characteristics, while progesterone, produced in the second phase of the menstrual cycle, supplements the action of oestrogen by thickening the lining of the uterus, ready for the possible implantation of a fertilised egg.

The Fallopian Tubes

The two fallopian tubes are each about 5 cm long, and extend from the sides of the uterus, passing upwards and outwards to end near each ovary. At the end of each fallopian tube are finger-like projections called **fimbrae** which encircle the ovaries.

Function

The function of the fallopian tubes is to convey the ovum from the ovary to the uterus. It is swept down the tube by peristaltic muscular contraction, assisted by the lining of ciliated epithelium.

Fertilisation of the ovum takes place within the fallopian tubes and it then passes to the uterus.

The Uterus

The uterus is a small hollow, pear shaped organ situated behind the bladder and in front of the rectum. It has thick muscular walls and is composed of three layers of tissue:

- The **perimetrium**: an outer covering which is part of the peritoneum (a serous membrane in the abdominal cavity). It covers the superior (top) part of the uterus.
- The **myometrium**: a middle layer of smooth muscle fibres. This layer forms 90 per cent of the uterine wall and is responsible for the powerful contractions that occur at the time of labour.
- The **endometrium**: a soft, spongy mucous membrane lining, the surface of which is shed each month during menstruation.

The uterus can be divided into three parts:

1 The **fundus** is the dome-shaped part of the uterus above the openings of the uterine tubes.
2 The **body** is the largest and main part of the uterus and leads to the cervix.
3 The **cervix** of the uterus is a thick fibrous muscular structure which opens into the vagina.

Function

The uterus is part of the female reproductive tract which is specialised to receive an ovum, and serves as the area in which an embryo grows and develops into a foetus. After puberty, the uterus goes through a regular cycle of changes which prepares it to receive, nourish and protect a fertilised ovum.

During pregnancy, the walls of the uterus relax to accommodate the growing foetus. If the ovum is not fertilised, the cycle ends with a short period of bleeding in which the endometrium undergoes periodic development and degeneration, known as the menstrual cycle.

The Vagina

The vagina is a 10–15cm muscular and elastic tube, lined with moist epithelium, which connects the internal organs of the female reproductive system with the external genitalia.

It is made up of vascular and erectile tissue and extends from the cervix (internally) of the uterus above to the vulva (externally) below.

Function

The function of the vagina is for the reception of the male sperm, and to provide a passageway for menstruation and for childbirth.

The wall of the vagina is sufficiently elastic to allow for expansion during childbirth. Between the phases of puberty and the menopause, the vagina also provides an acid environment, due to acid-secreting bacteria, in order to help prevent the growth of microbes that may infect the internal organs.

TASK 1 – THE FEMALE REPRODUCTIVE SYSTEM

Label the following parts of the female reproductive system:

ovaries   fallopian tubes   cervix   uterus   vagina

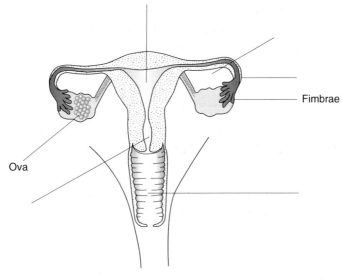

Fimbrae

Ova

Figure 69
Female reproductive organs

Puberty

This is the time at which the onset of sexual maturity occurs and the reproductive organs become functional. Changes in both sexes occur with the appearance of the secondary sexual characteristics, such as the deepening of the voice in a boy and growth of breasts in girls. These changes are brought about by an increase in sex hormone activity, due to stimulation of the ovaries and testes by the pituitary gonadotrophic hormones.

The Menstrual Cycle

Starting at puberty, the female reproductive system undergoes a regular sequence of monthly events, known as the menstrual cycle.

The ovaries undergo cyclical changes, in which a certain number of ovarian follicles develop. When one ovum completes the development process, it is released into one of the fallopian tubes. If fertilisation does not occur, the developed ovum disintegrates and a new cycle begins.

The menstrual cycle lasts approximately 28 days, although it can be longer or shorter than this. There are three stages of the menstrual cycle:

▶ profilterative (first) phase: days 7–14 of the cycle

- secretory (second) phase: days 14–28 of the cycle
- menstrual (third) phase: days 1–7 of the cycle.

Profilerative Phase

- At the beginning of the cycle an ovum develops within an ovarian follicle in the ovary. This is in response to a hormone released by the anterior lobe of the pituitary gland called the follicle stimulating hormone (FSH), which stimulates the follicles of the ovaries to produce the hormone oestrogen.
- Oestrogen stimulates the endometrium to promote the growth of new blood vessels and mucus-producing cells.
- When mature, it bursts from the follicle and travels along the fallopian tube to the uterus. This occurs about 14 days after the start of the cycle and is known as **ovulation**.

Secretory Phase

- A temporary endocrine gland, the corpus luteum, develops in the ruptured follicle in response to stimulation from the luteinising hormone (LH) secreted by the anterior lobe of the pituitary gland. The corpus luteum secretes the hormone progesterone, which together with oestrogen causes the lining of the uterus (endometrium) to become thicker and richly supplied with blood in preparation for pregnancy.
- After ovulation, the ovum can only be fertilised during the next 8–24 hours.
- If fertilisation does occur, the fertilised ovum becomes attached to the endometrium and the corpus luteum continues to secrete progesterone. Pregnancy then begins.
- The corpus luteum continues to secrete progesterone until the fourth month of pregnancy, by which time the placenta has taken over this function.

Menstrual Phase

- If the ovum is not fertilised, the cycle continues and the corpus luteum shrinks and the endometrium is shed. This is called **menstruation**.
- Over a period of about five days, the muscles of the wall of the uterus contract to expel the unfertilised egg, pieces of endometrial tissue and some tissue fluid.
- As soon as levels of progesterone drop (due to the breakdown of the endometrium and the corpus luteum), the pituitary gland starts producing progesterone again and hence stimulates the ovaries to produce another follicle and a new ovum. The cycle then begins again.

Pregnancy

Pregnancy takes approximately nine calendar months and is divided into three trimesters.

▶ The *first* trimester is the most important to the developing baby and is a time of radical hormonal changes. During this phase, all of the body systems develop.

▶ The *second* trimester consists of rapid fetal growth and completion of systemic development. Blood volume for the mother increases as additional workload is placed on all physiological functions. Cardiac output, breathing rate and urine production increase in response to fetal demands. The uterus enlarges greatly during pregnancy, along with the size of the breasts. Appetite increases in response to the fetal need for increasing amounts of nutrients.

▶ The *third* trimester is mostly a weight-gaining and maturing process, preparing the baby for life outside of the womb. Posture changes are evident at this stage as the mother gains more weight and internal organs are compressed. The body's connective tissue structure alters by softening, to allow for the expansion needed for the birth.

The Hormonal Changes that Occur during Pregnancy

During a typical menstrual cycle, the corpus luteum degenerates about two weeks after ovulation. Consequently, the levels of oestrogen and progesterone decline rapidly and the lining of the uterus (endometrium) is not maintained and is sloughed off as menstrual flow.

If this occurs after implantation the embryo becomes spontaneously aborted (miscarries). The mechanism that usually prevents this occurring involves a hormone called human chorionic gonadotrophic hormone (HCG), which is secreted by a layer of embryonic cells that surround the developing embryo. HCG causes the corpus luteum to be maintained in order to establish the pregnancy.

The maintenance of the corpus luteum is important for the first three months, after which the placenta is usually well developed and is able to secrete sufficient oestrogen and progesterone. The secretion of the hormones oestrogen and progesterone is important during pregnancy as they:

▶ maintain the uterine wall
▶ inhibit the secretion of the gonadotrophic hormones FSH and LH
▶ stimulate development of the mammary glands
▶ inhibit uterine contractions until birth
▶ cause enlargement of the reproductive organs

Menopause

After puberty, the menstrual cycle normally continues to occur at regular intervals into approximately the late 40s or early 50s (most commonly between 45 and 55). At this time, there are marked changes in which the cycle becomes increasingly irregular, until the cycle ceases altogether. This period in a woman's reproductive life is called the **menopause** (female climacteric).

During the menopause there is a change in the balance of the sex hormones; the ovaries cease responding to the follicle stimulating hormone (FSH), and this decline in function results in lower levels of oestrogen and progesterone secretion.

As a result of reduced oestrogen concentration and lack of progesterone, the female's secondary sexual characteristics undergo varying degrees of change, which may include a decrease in size of the vagina, uterus and uterine tubes, as well as atrophy of the breasts.

Other changes that occur commonly in response to low oestrogen concentration include increased loss of bone matrix, increasing the risk of osteoporosis, thinning of the skin and dryness of the mucous membrane lining the vagina.

Some women of menopausal age experience unpleasant vasomotor symptoms including sensations of heat in the face, neck and upper body (known as 'hot flushes'). Menopausal women may also experience varying degrees of headache, backache and fatigue, as well as emotional disturbances.

The Male Reproductive System

The male reproductive system consists of the:

▶ testes
▶ epididymides
▶ vas deferentia
▶ prostate gland
▶ ejaculatory ducts
▶ cowpers glands
▶ urethra
▶ penis.

The Testes

The testes are the reproductive glands of the male, and lie in the scrotal sac. Each testis consists of approximately 200 to 300 lobules; these are separated by connective tissue and filled with seminiferous tubules, in which sperm cells are formed. Between the tubules are a group of secretory cells known as the **interstitial** cells which produce male sex hormones.

The testes are specialised to produce and maintain sperm cells, and to produce male sex hormones known collectively as androgens. **Testosterone** is the most important androgen as it stimulates the development of the male reproductive organs. It is also responsible for the development and maintenance of the male secondary sexual characteristics.

Epididymides

The epididymides are coiled tubes leading from the seminiferous tubules of the testis to the vas deferens. They store and nourish immature sperm cells and promote their maturation until ejaculation.

Vas Deferentia

The vas deferentia are tubes leading from the epididymis to the urethra and through which the sperm are released.

KEY NOTE

The vas deferentia are cut in the operation known as a vasectomy, which produces sterilisation in the male.

Seminal Vesicles

The seminal vesicles are pouches lying on the posterior aspect of the bladder attached to the vas deferens. They secrete an alkaline fluid which contains nutrients and is added to sperm cells during ejaculation.

Ejaculatory Ducts

The two ejaculatory ducts are short tubes which join the seminal vesicles to the urethra.

Cowper's Glands

The cowper's glands are a pair of small glands that open into the urethra at the base of the penis. These glands produce further secretions to contribute to the seminal fluid, but less than that of the prostate gland or seminal vesicles.

Prostate Gland

The prostate gland is a male accessory gland about the size of a walnut. It lies in the pelvic cavity in front of the rectum and behind the symphysis pubis. During ejaculation it secretes a thin, milky fluid that enhances the mobility of sperm and neutralises semen and vaginal secretions.

KEY NOTE

The prostate gland commonly becomes enlarged in older men, causing difficulty in passing urine due to constriction of the urethra.

Urethra

The urethra provides a common pathway for the flow of urine and the secretion of semen. A sphincter muscle prevents both functions occurring at the same time.

Penis

The penis is composed of erectile tissue and is richly supplied with blood vessels. Its function is to convey urine and semen.

testes   epididymis   prostate gland   vas deferens

cowper's gland   urethra   penis

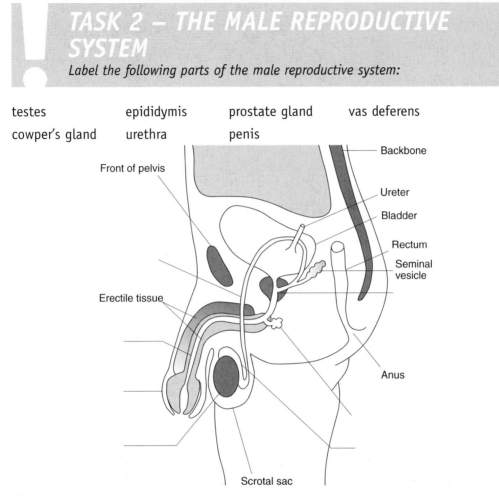

Figure 70

Male reproductive organs

Disorders of the Reproductive System

Female Disorders

Amenorhhea

Amenorrhea is the absence or stopping of the menstrual periods. Causes may be associated with deficiency of ovarian, pituitary or thyroid hormones, mental disturbances, depression, radical weight loss, stress, excessive exercise or a major change in surroundings or circumstances.

Dysmenorrhea

This condition is defined as painful and difficult menstruation. It presents with spasms and congestion of the uterus, resulting in cramping lower abdominal pains which start before or with the menstrual flow, and continue during menstruation. It is often associated with nausea, vomiting, headache and a feeling of faintness.

Endometriosis

Inflammation of the endometrium (the inner lining of the uterus). It presents with abnormal menstrual bleeding, lower abdominal pain and a foul-smelling discharge. Fever and malaise may accompany this condition.

Fibroid

An abnormal growth of fibrous and muscular tissue, one or more of which may develop in the muscular wall of the uterus. Fibroids can cause pain and excessive bleeding and become extremely large. Although they do not threaten life, they render pregnancy unlikely.
Some fibroids may be removed surgically; in other cases a hysterectomy may be necessary.

Pre-menstrual Syndrome

Pre-menstrual syndrome is a term for the physical and psychological symptoms experienced 3–14 days prior to the onset of menstruation.

The condition presents with varying symptoms: headache, bloatedness, water retention, backache, changes in coordination, abdominal pain, swollen and painful breasts, depression, irritability and craving for sweet things.

Infertility

The inability in a woman to conceive or in a man to induce conception. Female infertility may be due to a failure to ovulate, to obstruction of the fallopian tubes, or endometriosis.

Male Disorders

Prostatitis

Inflammation of the prostate gland, which is usually caused by bacteria. This condition presents with a frequency and urgency on passing urine (urine may be cloudy). High fever with chills, muscle and joint pain are common. A dull ache may be present in the lower back and pelvic area.

Infertility

Causes of male infertility can include decreased numbers or motility of sperm, or may be due to the total absence of sperm. In both male and female infertility, the cause may also be associated with stress.

SELF-ASSESSMENT QUESTIONS – THE REPRODUCTIVE SYSTEM

1. List the two functions of the ovaries.

2. What is the function of the fallopian tubes?

3. Where is the uterus situated anatomically, and what is its function?

4. What is the function of the vaginal secretions in a female?

5. List the functions of the testes.

6. Where in the male reproductive system are sperm cells

a) formed

b) stored to maturation?

7. What is the function of the prostate gland?

8. What is the collective term for male hormones?

CHAPTER 16

The Female Breast

> The female breasts are accessory organs to the female reproductive system, and their function is to produce and secrete milk after pregnancy.

A competent therapist needs to have a basic knowledge of the structure of the breast to understand how salon treatments may affect the size and shape of the breast.

By the end of this chapter you will be able to relate the following knowledge to your practical work carried out in the salon:

▶ brief structure of the breast
▶ the function of the breast
▶ factors affecting the size and shape of the breasts.

Anatomy of the Breast
Position

The breasts lie on the pectoral region of the front of the chest. They are situated between the sternum and the axilla, extending from approximately the second to the sixth rib. The breasts lie over the pectoralis major and serratus anterior muscles, and are attached to them by a layer of connective tissue.

Structure

The breasts consist of glandular tissue arranged in lobules, supported by connective, fibrous and adipose tissue. The lobes are divided into lobules which open up into milk ducts.

The milk ducts open into the surface of the breast at a projection called the nipple. Around each nipple, the skin is pigmented and forms the areola; this varies in colour from a deep pink to a light or dark brown colour. A considerable amount of fat or adipose tissue covers the surface of the gland and is found between the lobes. The skin on the breast is thinner and more translucent than the body skin.

Support

The breasts are supported and slung in powerful suspensory **cooper's ligaments**, which go around the breast, both ends being attached to the chest wall. The pectoralis major and serratus anterior muscles help to support the ligaments.

If the breast grows large due to adolescence or pregnancy, the cooper's ligaments may become irreparably stretched and the breast will then sag. With age, the supporting ligaments, along with the skin and the breast tissue, become thin and inelastic, and the breasts lose their support.

Physiology of the Breast

Lymphatic Drainage

The breasts contain many lymphatic vessels, and the lymph drainage is very extensive, draining mainly into the axillary nodes under the arms.

Blood Supply

The blood vessels supplying blood to the breast include the subclavian and axillary arteries.

Nerve Supply

There are numerous sensory nerve endings in the breast, especially around the nipple. When these touch receptors are stimulated in lactation, the impulses pass to the hypothalamus, and the flow of the hormone oxytocin is increased from the posterior lobe of the pituitary. This promotes the constant flow of milk when required.

Hormones

The hormones responsible for developing the breast are:

▶ **Oestrogen**: is responsible for the growth and development of the secondary sex characteristics
▶ **Progesterone**: causes the mammary glands to increase in size if fertilisation and subsequent pregnancy occurs.

Development of the Breasts

Puberty

The breast starts out as a nipple which projects from the surrounding ring of pigmented skin called the areola. Approximately two or three years before the onset of menstruation, the fat cells enlarge in response to the sex hormones (oestrogen and progesterone) released during adolescence.

KEY NOTE

The breasts change monthly in response to the menstrual cycle. The action of the female hormone progesterone increases blood flow to the breast which increases fluid retention, and the breast may increase in size, causing it to feel swollen and uncomfortable.

Pregnancy

During pregnancy, the increased production of oestrogen and progesterone causes an increase in blood flow to the breast. This causes an enlargement of the ducts and lobules of the breast in preparation for lactation, and there is an increase in fluid retention. The areola and the nipple enlarge and become more pigmented.

Menopause

The reduction in the female hormones during the menopause causes the glandular tissue in the breast to shrink, and the supporting ligaments, along with the skin, become thinner and lose their elasticity. Therefore, during the menopause the breasts begin to lose their support and uplift, although the degree of loss is dependent on the original strength of the suspensory ligaments.

Factors Determining Size and Shape

The size of the breast is largely determined by genetic factors, although there are other factors such as:

▶ amount of adipose tissue present
▶ fluid retention
▶ level of ovarian hormones in the blood and the sensitivity of the breasts to these hormones
▶ degree of ligamentary suspension
▶ exercise undertaken.

KEY NOTE

Exercise may help to strengthen the pectoral muscles which will help to support the ligaments and increase their uplift. However, if the wrong type of exercise is undertaken and insufficient support is not provided for the breasts during exercise, the ligaments may become irreparably stretched.

TASK 1 – THE STRUCTURE OF THE FEMALE BREAST

Label the parts of the female breast on Figure 71:

nipple   areola   pectoralis major muscle   lobules of glandular tissue

skin   cooper's ligaments   lactiferous (milk) ducts   connective tissue

adipose tissue

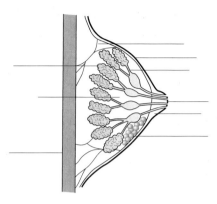

Figure 71
Cross section of the female breast

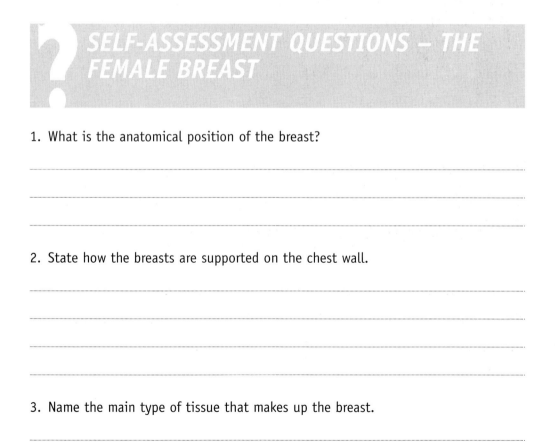

? SELF-ASSESSMENT QUESTIONS – THE FEMALE BREAST

1. What is the anatomical position of the breast?

....

....

....

2. State how the breasts are supported on the chest wall.

....

....

....

....

3. Name the main type of tissue that makes up the breast.

....

4. State the hormones responsible for the development of the breasts.

5. Briefly state the blood supply and venous drainage of the breast.

6. Briefly state the lymphatic drainage of the breast.

7. List the changes which occur to the breast during:

a) puberty

b) pregnancy

c) menopause

8. List four factors affecting the size and shape of the breasts.

....

....

....

....

CHAPTER 17

The Digestive System

In the digestive system, food is broken down and made soluble before it can be absorbed by the body for nutrition. Food is taken in through the mouth, broken into smaller particles and absorbed into the bloodstream, where it is utilised by the body. Waste materials not required by the body are then passed through the body to be eliminated. Once food has been absorbed by the body, it is converted into energy to fuel the body's activities; this is known as metabolism.

A competent therapist needs to know the process of the breakdown of food, in order to understand how the body utilises nutrients for efficient body functioning.

By the end of this chapter, you will be able to relate the following knowledge to your work carried out in the salon:

▶ the process of digestion from the ingestion of food to the elimination of waste
▶ the structure and functions of the organs associated with digestion
▶ the absorption of nutrients and their utilisation in the body
▶ sources and functions of the main food groups required for good health
▶ disorders of the digestive system.

The Structure and Function of Digestive Organs

Digestion occurs in the alimentary tract, which is a long continuous muscular tube, extending from the mouth to the anus. It consists of the following parts:

- the mouth
- the pharynx
- the oesophagus
- the stomach
- the small intestine
- the large intestine
- the anus.

The pancreas, gall bladder and the liver are accessory organs to digestion.

The Mouth

The digestive system commences in the mouth. Food is broken up into smaller pieces by the teeth, shaped into a ball by the tongue and mixed thoroughly with saliva from the salivary glands, which open into the mouth.

The smell and sight of food triggers the reflex action of the secretion of saliva in the mouth. Saliva enters the mouth from three pairs of salivary glands. These are the:

1 **sublingual** glands (located in the roof of the mouth under the tongue)
2 **submandibular** glands (located inside the arch of the mandible)
3 **parotid** glands (located superficial to a masseter muscle).

The enzyme 'salivary amylase' commences the digestion of starch or carbohydrates in the mouth.

KEY NOTE

An enzyme is a chemical catalyst which activates and speeds up a chemical reaction, without any change to itself. Enzymes are highly specific, in that each enzyme catalyses only one type of metabolic action. An example is salivary amylase, which will only act on starch, but has no effect on protein.

The Pharynx and Oesophagus

The ball of food is projected to the back of the mouth. The muscles of the pharynx force the food down the oesophagus, which is a long narrow tube linking the pharynx to the stomach. The food is then conveyed by peristalsis down the oesophagus to the stomach.

KEY NOTE

Peristalsis is the coordinated rhythmical contractions of the circular and oblique muscles in the wall of the alimentary canal. These muscles work in opposition to one another, to break food down and move it along the alimentary canal. Peristalsis is an automatic action stimulated by the presence of food, and occurs in all sections of the alimentary canal.

The Stomach

The stomach is a curved muscular organ, positioned in the left hand side of the abdominal cavity, below the diaphragm. The stomach has:

- a serous membrane which prevents friction
- a muscular coat which assists the mechanical breakdown of food
- numerous gastric glands which secrete gastric juice
- a mucous coat, secreting mucous to protect the stomach lining from the damaging effects of the acidic gastric juice.

The functions of the stomach are to:

- churn and break up large particles of food mechanically
- mix food with gastric juice to begin the chemical breakdown of food
- commence the digestion of *protein*.

The main constituents of gastric juice, produced and secreted by cells in the stomach wall are:

1 **Pepsin,** which is secreted as pepsinogen by the chief cells. Pepsinogen is an enzyme which starts the breakdown of proteins.
2 **Hydrochloric acid** is manufactured by the oxyntic cells. It provides the acidic conditions needed for pepsin to become active, to kill germs present in food, and to prepare it for intestinal digestion.
3 **Mucus**, which is secreted by the neck cells. In addition to its lubricative action, mucus lines the stomach wall to protect the stomach lining from the damaging effects of the acidic gastric juice.

Food stays in the stomach for approximately five hours, until it has been churned down to a liquid state called **chyme**. Chyme is then released at intervals into the first part of the small intestine.

The Small Intestine

The small intestine is made up of the same coats as the stomach and consists of three parts:

1 the duodenum (the first part of the small intestine)
2 the jejunum
3 the ileum (where the main absorption of food takes place).

Special features of the small intestine are the thousands of minute projections called **villi**, into which the nutrients pass to be absorbed into the bloodstream. The muscles in the wall of the small intestine continue the mechanical breakdown of food by the action of peristalsis. The chemical breakdown of food is completed by the following juices, which prepare the food to be absorbed into the bloodstream:

Bile

This is stored by the gall bladder, a muscular, membranous bag situated on the underside of the right lobe of the liver. Bile is an alkaline liquid consisting of water, mucus, bile pigments, bile salts and cholesterol, and is released at intervals from its duct when food enters the duodenum. The function of bile is to neutralise the chyme and break up any fat droplets in a process called **emulsification.**

Pancreatic Juice

This is produced by the pancreas, a gland extending from the loop of the duodenum to behind the stomach. The pancreas secretes pancreatic juice into the duodenum, and the enzymes contained within it continue the digestion of protein, carbohydrates and fat.

The pancreas also has an endocrine function in that it secretes insulin from the islet of langerhans cells in the pancreas. Insulin is important to digestion because it regulates blood sugar levels.

- **Trypsin** continues the breakdown of proteins
- **Pancreatic Amylase** continues the breakdown of starch and has the same effects as salivary amylase
- **Pancreatic Lipase** breaks down lipids into fatty acids and glycerol.

Intestinal Juice

This is released by the glands of the small intestine and completes the final breakdown of nutrients, including simple sugars to glucose and protein to **amino acids.**

Carbohydrate digestion is completed by the following enzymes:

- **Maltase** splits maltose into glucose
- **Sucrase** splits sucrose into glucose and fructose
- **Lactase** splits lactose into glucose and galactose.

Protein digestion is completed by **peptidases** which split short chain **polypeptides** into amino acids.

Absorption of the Digested Food

The absorption of the digested food occurs by diffusion through the villi of the small intestine. The villi are well supplied with blood capillaries to allow the digested food to enter.

- Simple sugars from *carbohydrate* digestion and amino acids from *protein* digestion pass into the bloodstream via the villi and are then carried to the liver to be processed.
- Products of *fat* digestion pass into the intestinal lymphatics, which absorb the fat molecules and carry them through the lymphatic system before they reach the blood circulation.
- *Vitamins* and *minerals* travel across to the blood capillaries of the villi and are absorbed into the bloodstream, to assist in normal body functioning and cell metabolism.

The Liver

The liver is the largest gland in the body, and is situated in the upper right hand side of the abdominal cavity, under the diaphragm. It has a soft reddish-brown colour and four lobes. Its internal structure is made up of cells called **hepatocytes**. The liver receives oxygenated blood from the hepatic artery, and deoxygenated blood from the hepatic portal vein. Blood from the digestive tract which is carried in the portal veins brings newly absorbed nutrients into the sinusoids, and nourishes the liver cells. The liver is a vital organ and therefore has many important functions in the metabolism of food; it regulates the nutrients absorbed from the small intestine to make them suitable for use in the body's tissues. Its functions are:

- **Secretion of bile** – bile is manufactured by the liver but is stored and released by the gall bladder to assist the body in the breakdown of fats.
- **Regulation of blood sugar levels** – when the blood sugar levels rise after a meal, the liver cells store excess glucose as glycogen. Some glucose may be stored in the muscle cells as muscle glycogen.

▶ When both these stores are full, surplus glucose is converted into fat by the liver cells.

▶ **Regulation of amino acid levels** – as our bodies cannot store excess protein and amino acids, they are processed by the liver; some are removed by the liver cells and are used to make plasma proteins, some are left for the body cells tissues' use, whilst the rest are deaminated and excreted as urea in the kidneys.

▶ **Regulation of the fat content of blood** – the liver is involved in the processing and transporting of fats; those already absorbed in the diet are used for energy, and excess fats are stored in the tissues.

▶ **Regulation of plasma proteins** – the liver is active in the breakdown of worn out red blood cells.

▶ **Detoxification** – the liver detoxifies harmful toxic waste and drugs, and excretes them in bile or through the kidneys.

▶ **Storage** – the liver stores vitamins A, D, E, K and B12, and the minerals iron, potassium and copper.

▶ The liver can also hold up to a litre of blood. During exercise, the liver supplies extra blood and increases oxygen transport to the muscles.

The Production of Heat

Due to its many functions, the liver generates heat. Once all the nutrients have been absorbed into the bloodstream, they are transported to the body's cells for metabolism:

▶ **Glucose**, which is the end product of carbohydrate digestion and is used to provide energy for the cells to function.

▶ **Amino acids**, which are the end products of protein digestion. They are used to produce new tissues, repair damaged cell parts and to formulate enzymes, plasma proteins and hormones.

▶ **Fatty acids and glycerol** are the end products of fat digestion. Fats are used primarily to provide heat and energy, in addition to glucose. Those fats which are not required immediately by the body are used to build cell membranes, and some are stored under the skin or around vital organs such as the kidneys and the heart.

When all the body's nutrients have been assimilated by the body, the fate of the undigested food is to pass into the large intestine where it is eventually eliminated from the body.

The Large Intestine

The large intestine coils around the small intestine, and is made up of bands of longitudinal muscle and folds of mucosa which secrete mucous.

The colon is the main part of the large intestine and is divided into four sections: ascending, transverse, descending and sigmoid colons:

1 The **ascending colon** is the part that passes upwards on the right side of the abdomen from the caecum (a pouch at the junction of the small and large intestines) to the lower edge of the liver.
2 The **transverse colon** is the longest and most mobile part, and extends across the abdomen from right to left below the stomach.
3 The **descending colon** is the part that passes downwards along the left side of the abdominal cavity to the brim of the pelvis.
4 The **sigmoid colon** is the S-shaped part of the large intestine, between the descending colon and the rectum.

The functions of the large intestine are:

▶ absorption of most of the water from the faeces, in order to conserve moisture in the body
▶ formation and storage of faeces (which consists of undigested food, dead cells and bacteria)
▶ production of mucus to lubricate the passage of faeces
▶ the expulsion of faeces out of the body, through the anus.

The rectum is the last part of the large intestine. It is about 12 cm long and runs from the sigmoid colon to the anal canal. It is firmly attached to the sacrum and ends about five centimetres below the tip of the coccyx, where it becomes the anal canal. Faeces are stored in the rectum before defecation.

The Anus

The anus is an opening at the lower end of the alimentary canal (the anal canal), through which faeces is discharged. The anus is guarded by two sphincter muscles:

1 an *internal* sphincter composed of smooth muscle under *involuntary* control
2 an *external* sphincter composed of skeletal muscle under *voluntary* control.

The anus remains closed, except during defecation.

TASK 1 – MAIN DIGESTIVE ORGANS
Label the parts of the digestive system on Figure 72:

mouth   stomach   liver   pancreas   pharynx

oesophagus   gall bladder   small intestine ascending colon   transverse colon

descending colon   rectum   anus

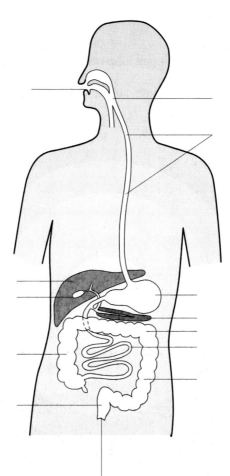

Figure 72
Main digestive organs

TASK 2 – THE PROCESS OF DIGESTION

Select from the following words to fill in the gapped sentences:

▶ small intestine, colon, bile, simple sugars, villi, mouth, gastric juice, carbohydrates, liver, water, vitamins, pancreatic, anus, saliva, oesophagus, chyme, rectum, intestinal, amino acids, faeces, peristalis.

Digestion commences in the _____, where food is chewed by the teeth and mixed thoroughly with _____, which contains an enzyme called amylase which starts to digest _____.

Food then passes down the _____ to the stomach. The food is then conveyed by a process of rhythmic muscular contractions called _____.

In the stomach the food is churned up and mixed with _____ which contains enzymes to digest protein.

The food stays in the stomach for approximately five hours until it has been churned down into a liquid state called _____.

Food is then passed into _____ _____ where more enzymes continue the chemical breakdown of food.

The food is also mixed with _____ which is manufactured in the liver to help emulsify fat; _____ juice from the pancreas to continue the digestion of protein, carbohydrates, and _____ juice which completes the final breakdown of nutrients, including simple sugars to glucose and protein to amino acids.

The absorption of the digested food occurs by diffusion through the _____ of the small intestine, which are small finger like projections well supplied with blood capillaries.

_____ _____ from carbohydrate digestion, _____ _____ from protein digestion, _____ and minerals are absorbed in the blood capillaries and products of fat digestion are absorbed into the intestinal lymphatics. The capillaries join to form the hepatic portal vein which transports the digested food to the _____ to be regulated before being utilised by the body's tissues. The undigested food passes into the _____ where a large amount of _____ is absorbed. The solid undigested matter, known as _____ passes into the _____ where it is stored before being passed out of the body through the _____.

Nutrition

Nutrition is:

▶ the utilisation of food to facilitate growth and to maintain the normal working of the body.

Poor nutrition can have a dramatic effect on our general health, energy levels, sleep patterns and stress response. Nutritional problems often result from not following dietary recommendations. This is often due to the fact that nutrition involves more than just eating. Eating affects the mood and therefore the issue is often an emotional topic.

An ideal diet is said to be one which is low in fats and sugars, moderately low in protein and dairy products, with the bulk of the calories coming from carbohydrates. Cycles of high and low blood sugar are as a result of an unbalanced diet, which is aggravated by improper pacing between meals and consuming stimulants such as tea, coffee, alcohol and fizzy drinks such as cola.

It is important therefore for a therapist to have a foundation knowledge of basic food groups, to be able to understand their role in body functioning and to advise clients correctly.

Carbohydrates

This group is also known as starches and sugars.

▶ **Dietary sources** – bread, cereals, potatoes, fruit and sugars.
▶ **Main functions** – the body's main source of energy. Required for the metabolism of other nutrients such as proteins and fats.

Proteins

▶ **Dietary sources** – first class proteins: fish, milk, eggs, meat. Second class: pulses, beans, peas.
▶ **Main functions** – growth and repair of the body tissues. Used in the production of hormones and enzymes.

Fats

Fats are classified as saturated or unsaturated, depending on whether they are solid (saturated) or liquid (unsaturated) at room temperature.

▶ **Dietary sources** – meat, milk, cheese, butter, eggs.
▶ **Main functions** – a source of stored energy. Offer support and protection for the body, and used to build cell structures.

Water

Although water is not usually considered as food, it is nevertheless an essential nutrient needed by every part of the body. Because our bodies are more than 70 per cent water, it is recommended that at least 6–8 glasses of fresh water are consumed every day for efficient body functioning.

▶ **Dietary sources** – fresh water, fruit and vegetables.
▶ **Main functions** – aids digestion and elimination. Is essential to maintaining the body's fluid balance. Aids in the transport of substances around the body.

Fibre

Although fibre is not broken down into nutrients, it is a very necessary component for effective digestion.

▶ **Dietary sources** – pulses, peas, beans, brown rice, wholemeal bread, jacket potatoes, green leafy vegetables.
▶ **Main functions** – aids digestion and bowel functioning. Provides the bulk in food to satisfy the appetite.

Vitamins

Vitamins are essential for normal physiological and metabolic functioning of the body. They regulate the body's processes and contribute to its resistance to disease. Vitamins are divided into two groups, according to whether they are soluble in water or fat.

Vitamin A (fat soluble)
▶ **Dietary sources** – carotene in carrots, liver, kidney, eggs, dairy products, fish and liver oils.
▶ **Main functions** – essential for healthy vision, healthy skin and mucous membrane.

Vitamin D (fat soluble)
▶ **Dietary sources** – fish liver oils, fatty fish, margarine, eggs. Is also synthesised from ultra violet light.
▶ **Main functions** – essential for healthy teeth and bones. Maintains the blood calcium level by increasing calcium absorption from food.

Vitamin E (fat soluble)
▶ **Dietary sources** – peanuts, wheatgerm, milk, butter, eggs.

▶ **Main functions** – inhibits the oxidation of fatty acids that help form cell membranes.

Vitamin K (fat soluble)

▶ **Dietary sources** – green leafy vegetables, cereals, liver, fruit.
▶ **Main functions** – essential for blood clotting.

Vitamin B1 (water soluble)

▶ **Dietary sources** – egg yolk, liver, milk, wholegrain cereals, vegetables, fruit.
▶ **Main functions** – necessary for the steady release of energy from glucose.

Vitamin B2 (water soluble)

▶ **Dietary sources** – milk, liver, eggs, yeast.
▶ **Main functions** – essential for using energy released from food.

Vitamin B5 (water soluble)

▶ **Dietary sources** – wholegrain cereals, yeast extract, liver, beans, nuts and meat.
▶ **Main functions** – involved in the breakdown of glucose to release energy.

Vitamin B6 (water soluble)

▶ **Dietary sources** – wholegrain cereals, yeast extract, liver, meat, nuts, bananas, salmon, tomatoes.
▶ **Main functions** – necessary for the metabolism of protein and fat.

Vitamin B12 (water soluble)

▶ **Dietary sources** – liver, kidney, milk, eggs, cheese.
▶ **Main functions** – necessary for the formation of red blood cells in bone marrow. Is also involved in protein metabolism.

Folic Acid (water soluble)

▶ **Dietary sources** – liver, kidney, fresh leafy vegetables, oranges, bananas.
▶ **Main functions** – essential for normal production of red and white blood cells.

Vitamin C (water soluble)

▶ **Dietary sources** – citrus fruits and blackcurrants.
▶ **Main functions** – assists in the formation of connective tissue and collagen. Helps prevent bleeding and aids healing.

Minerals

Minerals provide the body with materials for growth and repair, and for the regulation of body processes. They are needed in trace amounts and are used to build bone, to work muscles, to support various organs and to transport oxygen and carbon dioxide.

Calcium

- **Dietary sources** – milk, egg yolk, cheese, green leafy vegetables.
- **Main functions** – essential for the formation of healthy bones and teeth, blood coagulation, and the normal function of muscles and nerves.

Iron

- **Dietary sources** – liver, kidney, red meats, egg yolk, nuts, green vegetables.
- **Main functions** – essential for the production of haemoglobin in red blood cells.

Phosphorus

- **Dietary sources** – cheese, eggs, white fish, wholemeal bread, peanuts, yeast extract.
- **Main functions** – important in the formation of bones and teeth, muscle contraction and the transmission of nerve impulses.

Sulphur

- **Dietary sources** – egg yolk, fish, red meat, liver.
- **Main functions** – is the main component of structural proteins (those in the skin and hair).

Sodium and Chlorine

- **Dietary sources** – table salt, bacon, kippers. Is found in all body fluids.
- **Main functions** – maintains fluid balance in the body. Is necessary for the transmission of nerve impulses and contraction of muscle.

Magnesium

- **Dietary sources** – green vegetables and salad.
- **Main functions** – important for the formation of bone. Is required for the normal functioning of muscles and nerves.

Disorders of the Digestive System

Anorexia Nervosa

A psychological illness in which clients starve themselves or use other techniques such as vomiting or laxatives, to induce weight loss. They are motivated by a false perception of their body image and a phobia of becoming fat. The result is a severe loss of weight, with amenorrhea and even death from starvation.

Bulimia

A psychological illness which is characterised by overeating (bingeing), followed by self-induced vomiting.

Colitis

Inflammation of the colon. The usual symptoms are diarrhoea, sometimes with blood and mucus, and lower abdominal pain.

Constipation

A condition where there is difficulty in passing stools or where there is infrequent evacuation of the bowel. Causes may be dietary due to reduced fibre and fluid intake, certain medications or intestinal obstruction.

Diarrhoea

A condition where there is frequent bowel evacuation or the passage of abnormally soft or liquid faeces. It may be caused by intestinal infections or other forms of intestinal inflammation such as colitis or irritable bowel syndrome (see below).

Gall Stones

A hard pebble-like mass which is formed within the gall bladder. The condition may be asymptomatic, or indigestion and colicky pain may be present.

Changes in the composition of bile cause cholesterol and/or bile pigment bilirubin to form stones. Stagnation of bile and inflammation of the gall bladder increase the concentration of bile and promote stone formation.

Haemorrhoids

A condition with abnormal dilatation of veins in the rectum. It is caused by increased pressure in the venous network of the rectum. If the haemorrhoids are chronic, they may be seen or felt as soft swellings in the anus.

Hepatitis

Inflammation of the liver caused by viruses, toxic substances or immunological abnormalities.

▶ **Hepatitis A** is highly contagious and is transmitted by the faecal–oral route. It is transmitted by ingestion of contaminated food, water or milk. The incubation period is 15–45 days.
▶ **Hepatitis B**, also known as serum hepatitis, is more serious than Hepatitis A. It lasts longer and can lead to cirrhosis, cancer of the liver and a carrier state. It has a long incubation period of 1.5–2 months. The symptoms may last from weeks to months. The virus is usually transmitted through infected blood, serum or plasma; however it can spread by oral or sexual contact as it is present in most body secretions.
▶ **Hepatitis C** can cause acute or chronic hepatitis, and can also lead to a carrier state and liver cancer. It is transmitted through blood transfusions or exposure to blood products.

Most clients with hepatitis are jaundiced, but they can appear to be entirely healthy.

Note: Hepatitis as a side effect of drugs and alcohol intake is **not** infective.

Hernia

An abnormal protrusion of an organ or part of an organ through the wall of the body cavity in which it normally lies.

Hiatus Hernia

This is the most common type of hernia, and occurs when part of the stomach is protruding into the chest. This sometimes causes no symptoms at all, but it can cause acid reflux, when acid from the stomach passes to the oesophagus, causing pain and heartburn.

Irritable Bowel Syndrome

A common condition in which there is recurrent abdominal pain with constipation and / or diarrhoea, and bloating. Clients with stress and hectic lifestyles are more vulnerable to this illness. They usually defecate infrequently, usually in the morning, but may feel that their bowel is not empty, or they may pass stool-like pellets.

Ulcers

A break in the skin or a break in the lining of the alimentary tract which fails to heal and is accompanied by inflammation.

Peptic, duodenal and gastric ulcers can present with increased acidity, epigastric pain and heartburn. This may be worst when hungry or after consumption of irritating foods and alcohol, for example spicy or fatty foods, mayonnaise, wines and spirits. It can present with similar symptoms of a hiatus hernia and reflux.

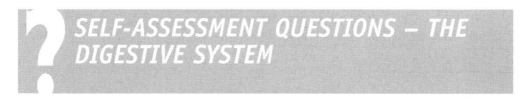

SELF-ASSESSMENT QUESTIONS – THE DIGESTIVE SYSTEM

1. What is meant by the term 'metabolism'?

2. Define the term 'enzyme'.

3. State where the following enzymes are produced. Briefly state their effects

a) salivary amylase

b) pepsin

c) trypsin

d) pancreatic lipase

4. What is meant by the term 'peristalsis' in connection with digestion? Where does this occur?

5. State the principal function/s of the following digestive organs:

a) mouth

b) stomach

c) gall bladder

d) small intestine

e) large intestine

6. List four metabolic functions of the liver.

7. Briefly explain how the following nutrients are utilised by the body:

a) glucose

b) amino acids

c) fats

CHAPTER 18

The Urinary System

The kidneys and associated structures of the urinary system are all part of the excretory system, along with the skin, lungs and the intestines.

The urinary system is made up of excretory organs which are involved in the processing and elimination of normal metabolic waste from the body. Waste products such as urea and uric acid, along with excess water and mineral salts must be removed from the body in order to maintain good health. If these waste materials were allowed to accumulate in the body they would cause ill health.

The primary function of the urinary system, therefore, is to regulate the composition and the volume of body fluids, in order to provide a constant internal environment for the body.

A competent therapist needs to be able to have knowledge of the outline of the urinary system to understand how fluid balance is controlled in the body.

By the end of this chapter, you will be able to relate the following knowledge to your practical work carried out in the salon:

▶ the structure of the individual parts of the urinary system
▶ the functions of the individual parts of the urinary system
▶ the regulation of fluid balance in the body
▶ disorders of the urinary system.

Structure of the Urinary System

The urinary system consists of the following parts:

▶ two **kidneys** which secrete urine
▶ two **ureters** which transport urine from the kidneys to the bladder
▶ one **urinary bladder** where urine collects and is temporarily stored
▶ one **urethra** through which urine is discharged from the bladder and out of the body.

The Kidneys

The kidneys lie on the posterior wall of the abdomen, on either side of the spine, between the level of the twelfth thoracic vertebrae and the third lumbar vertebrae. A kidney has an outer fibrous capsule and is supported by adipose tissue. It has two main parts:

1 the **outer cortex**, which is reddish-brown, and is the part where fluid is filtered from blood.
2 the **inner medulla**, which is paler in colour and is made up of conical-shaped sections called renal pyramids. This is the area where some materials are selectively re-absorbed back into the bloodstream.

How the Kidneys Work

The cortex and the medulla contain tiny blood filtration units called **nephrons**. A single kidney has more than a million nephrons. The medial border of the kidney is called the **hilus**, and is the area where the renal blood vessels leave and enter the kidney.

The blood that needs to be processed enters the medulla from the renal artery. Inside the kidney, the renal artery splits into a network of capillaries called the **glomerulus**, which filter the waste. Almost encasing the glomerulus lies a sac called the **bowmans capsule**.

The blood pressure in the glomerulus is maintained at a high level, assisted by the fact that the arteriole feeding into the glomerulus has a larger diameter than the arteriole leaving it. This pressure forces fluid out through the walls of the glomerulus, together with some of the substances of small molecular size able to pass through the capillary walls into the bowmans capsule.

This process constitutes simple filtration. The filtered liquid continues through the convoluted tubules and the loop of henle, which are surrounded by capillaries. Some substances contained within the waste such as glucose, amino acids, mineral salts and vitamins are re-absorbed back into the bloodstream, as the body cannot afford to lose them.

Excess water, salts and the waste product urea are all filtered and processed through the kidneys and the treated blood leaves the kidney via the renal vein. Urine, the waste product of filtration produced by the kidney, collects in a funnel-shaped structure called the renal pelvis, from which it flows into the ureter.

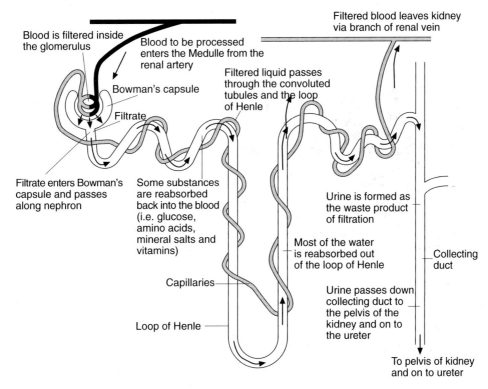

Figure 73
A nephron (blood filtration unit inside a kidney)

KEY NOTE

Urine consists of water, salts and protein wastes, and varies in its colour according to its composition and quantity. An analysis of the substances present in the urine is a good indication of the state of health of the body, and urine tests are often used to diagnose disorder and diseases in the body.

Function

The functions of the kidney are:

- filtration of impurities and metabolic waste from blood, and preventing poisons from fatally accumulating in the body
- regulation of water and salt balance in the body
- maintenance of the normal pH balance of blood
- formation of urine.

The Role of the Kidneys in Fluid Balance

The amount of fluid taken into the body must equal the amount of fluid excreted from it, in order for the body to maintain a constant internal environment. The balance between water intake and water output is controlled by the kidneys.

- **Water intake** – water is mainly taken into the body as liquid through the process of digestion; however, some is also released through the cells' metabolic activities.
- **Water output** – water is lost from the body in the following ways:

1 through the kidneys as urine
2 through the alimentary tract as faeces
3 through the skin as sweat
4 through the lungs as saturated exhaled breath.

The kidneys are responsible for regulating the amount of water contained within the blood:

- If you have an excess of water in the blood, the blood concentration will be dilute, and the nerve receptors in the hypothalamus will trigger the pituitary gland to send a message to the kidneys to reduce water reabsorption, in order that a more dilute urine is eliminated from the body.
  This mechanism *reduces* the amount of water in the blood back to an acceptable level.

- If your blood is too concentrated and you do not have enough water, the nerve receptors in the hypothalamus trigger the pituitary gland to send a message to the kidneys to increase water re-absorption, and a more concentrated urine is produced.
  This *increases* the water content of blood back to an acceptable level and helps to conserve water in the body.

Factors affecting fluid balance in the body include:

▶ **body temperature**: if the body temperature increases, more water is lost from the body in sweat.
▶ **diet**: a high salt intake can result in increased water re-absorption which reduces the volume of urine produced. Diuretics such as alcohol, tea and coffee can also increase the volume of urine.
▶ **emotions**: nervousness can result in an increased production of urine.
▶ **blood pressure**: when the blood pressure inside the kidney tubules rises, less water is re-absorbed and the volume of urine will be *increased* (and not decreased). When the blood pressure inside the kidney tubules falls, more water is re-absorbed into the blood and the volume of urine will be *decreased*.

The Ureters

The ureters are two very fine muscular tubes which transport urine from the pelvis of the kidney to the urinary bladder. They consist of three layers of tissue:

1 an outer layer of fibrous tissue
2 a middle layer of smooth muscles
3 an inner layer of mucous membrane.

Function

Their function is to propel urine from the kidneys into the bladder by the peristaltic contraction of their muscular walls.

The Urinary Bladder

This is a pear-shaped sac which lies in the pelvic cavity, behind the symphysis pubis. The size of the bladder varies according to the amount of urine it contains. The bladder is composed of four layers of tissue:

1 a serous membrane which covers the upper surface

2 a layer of smooth muscular fibres

3 a layer of adipose tissue

4 an inner lining of mucous membrane.

Functions

▶ The storage of urine.

▶ It expels urine out of the body, assisted by the muscular wall of the bladder, the lowering of the diaphragm and the contraction of the abdominal cavity.

The Urethra

This is a canal which extends from the neck of the bladder to the outside of the body. The length of the urethra differs in males and females; the female urethra being approximately only four centimetres in length, whereas the male urethra is longer, at approximately 18 to 20 centimetres in length. The exit from the bladder is guarded by a round sphincter of muscles which must relax before urine can be expelled from the body.

The urethra is composed of three layers of tissue:

1 a muscular coat, continuous with that of the bladder

2 a thin spongy coat, which contains a large number of blood vessels

3 a lining of mucous membrane.

Function

▶ It serves as a tube through which urine is discharged from the bladder to the exterior.

▶ As the urethra is longer in a male, it also serves as a conducting channel for semen.

Disorders of the Urinary System

Cystitis

An inflammation of the urinary bladder, usually caused by infection of the bladder lining. Common symptoms are pain just above the pubic bone, lower back or inner thigh, blood in the urine, and frequent, urgent and painful urination.

Incontinence

This is a condition in which the individual is unable to control urination voluntarily. Loss of muscle tone and problems with innervation are associated with this condition.

Kidney Stones

Deposits of substances found in the urine which form solid stones with the renal pelvis of the kidney, the ureter or the bladder. This condition can be extremely painful. Stones are usually removed by surgery.

Urinary Tract Infection

A bacterial infection of one or more of the structures of the urinary system. Symptoms include fever, lower back pain, frequency of urination, a burning sensation on passing urine (urine may be blood stained and cloudy).

TASK I – MAIN URINARY ORGANS

Label the main urinary organs on Figure 74:

kidney   urethra   bladder   ureter

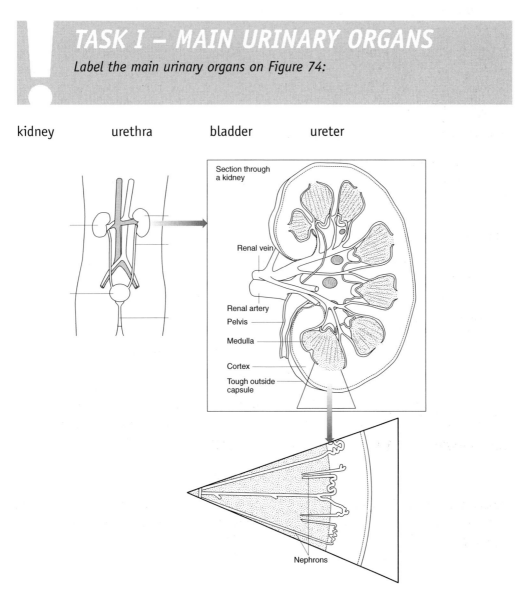

Section through a kidney

Renal vein

Renal artery

Pelvis

Medulla

Cortex

Tough outside capsule

Nephrons

Figure 74

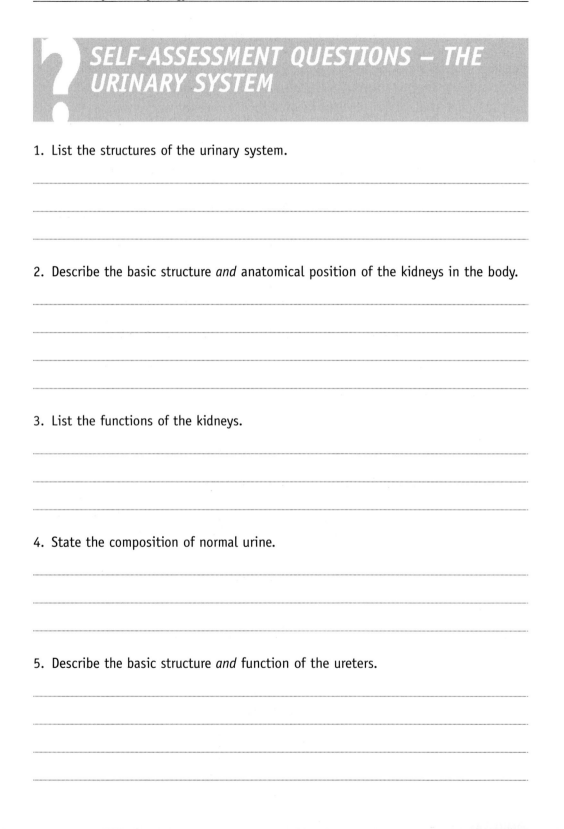

SELF-ASSESSMENT QUESTIONS – THE URINARY SYSTEM

1. List the structures of the urinary system.

2. Describe the basic structure *and* anatomical position of the kidneys in the body.

3. List the functions of the kidneys.

4. State the composition of normal urine.

5. Describe the basic structure *and* function of the ureters.

6. Describe the basic structure *and* function of the urinary bladder.

7. Describe the basic structure *and* function of the urethra.

8. Briefly describe the role of the kidneys in fluid balance in the body.

9. List two factors affecting fluid balance in the body.

10. Name two excretory organs other than the kidneys.

Candidate Competency Record

Anatomy and Physiology – Therapy Basics

Chapter 1 – Cells and Tissues	Date Competency Achieved	Assessor's Initials and No:
Task 1: The Structure of a Cell		
Task 2: Types of Tissue		
Self-Assessment Questions		
Chapter 2 – The Skin		
Task 1: The Appendages of the Skin		
Task 2: The Structure of the Skin		
Task 3: Skin Diseases and Disorders		
Self-Assessment Questions		
Chapter 3 – The Hair		
Task 1: The Structure of a Hair		
Task 2: The Hair in its Follicle		
Self-Assessment Questions		
Chapter 3 – The Nail		
Task 1: The Structure of a Nail		
Task 2: Cross Section of the Nail		
Task 3: Common Nail Diseases		

Self-Assessment Questions		
Chapter 5 – The Skeletal System		
Task 1: Major Bones of the Skull and Face		
Task 2: Bones of the Neck and Shoulder		
Task 3: Primary Bones of the Skeleton		
Self-Assessment Questions		
Chapter 6 – Joints		
Task 1: The Structure of a Synovial Joint		
Task 2: Types of Synovial Joint		
Self-Assessment Questions		
Chapter 7 – The Muscular System		
Task 1: Muscles of the Head and Neck		
Self-Assessment Questions		
Task 2: Muscles of the Shoulders		
Task 3: Muscles of the Upper Limbs		
Task 4: Muscles of the Lower Limbs		
Task 5: Muscles of Anterior of Trunk		
Task 6: Muscles of Respiration		
Task 7: Muscles of Posterior of Trunk		

Self-Assessment Questions		
Chapter 8 The Blood Circulatory System		
Task 1: Structural and Functional Differences between Arteries, Veins and Capillaries		
Task 2: The Structure of the Heart		
Self-Assessment Questions		
Chapter 9 – The Lymphatic System		
Task 1: Lymph Nodes of the Head and Neck		
Task 2: Lymph Nodes of the Body		
Self-Assessment Questions		
Chapter 10 – The Immune System		
Self-Assessment Questions		
Chapter 11 – The Respiratory System		
Task 1: Structures of the Respiratory System		
Self-Assessment Questions		
Chapter 12 – The Olfactory System		
Self-Assessment Questions		
Chapter 13 – The Nervous System		
Task 1: The Principal parts of the Brain		

Task 2: Effects of the Sympathetic and Parasympathetic Nervous Systems		
Self-Assessment Questions		
Chapter 14 – The Endocrine System		
Task 1: Main Endocrine Glands		
Task 2: Hormonal Secretions of the main Endocrine Glands		
Self-Assessment Questions		
Chapter 15 – The Reproductive System		
Task 1: The Female Reproductive System		
Task 2: The Male Reproductive System		
Self-Assessment Questions		
Chapter 16 – The Female Breast		
Task 1: The Structure of the Female Breast		
Self-Assessment Questions		
Chapter 17 – The Digestive System		
Task 1: Main Digestive Organs		
Task 2: The Process of Digestion		
Self-Assessment Questions		
Chapter 18 – The Urinary System		
Task 1: Main Urinary Organs		
Self-Assessment Questions		

Bibliography

Beazley Mitchell (1976) *The Atlas of Body and Mind,* Mitchell Beazley Publishers Ltd

Beckett, B.S. (1990) *Illustrated Human and Social Biology*, Oxford University Press.

Bennett R. (1995) *The Science of Beauty Therapy*, London: Hodder & Stoughton

Bird Mary SRN RN FAMS (1999*) Medical Terminology and Clinical Procedures 2nd Edition,* Magister Consulting Ltd

Fritz Sandy, Maison Paholsky Kathleen, Grosenbach M. James (1999) *Mosby's Basic Science for Soft Tissue and Movement Therapies,* Mosby

Gaudin, A.J. and Jones, K.C. (1989) *Human Anatomy and Physiology,* San Diego, CA: Harcourt Brace Jovanovich

Hole, J.W. Jr (1993) *Human Anatomy and Physiology*, Wm. C. Brown

Market House Books Ltd (1998) *Oxford Concise Colour Medical Dictionary,* Oxford University Press

Moffatt, D.B. and Mottram, R.F. (1983) *Anatomy and Physiology for Physiotherapists*, Oxford; Blackwell Scientific Publications

Premkumar Kalyani (1996) *Pathology A to Z,* Van Pub Books (ISBN 0 9680730 0 X)

Rowett, H.G Q. (1999) *Basic Anatomy and Physiology*, London, John Murray

Ross & Wilson, Ann Waugh and Allison Grant, (2001) *Anatomy & Physiology in Health and Illness*, Edinburgh: Churchill Livingstone

Wright, D. (1983) *Human Biology*, Oxford: Heinemann Educational

INDEX